Renewable Energy: Research, Development and Policies

Renewable Electric Power Distribution Engineering

Renewable Energy: Research, Development and Policies

Additional books and e-books in this series can be found on Nova's website under the Series tab.

RENEWABLE ENERGY: RESEARCH, DEVELOPMENT AND POLICIES

RENEWABLE ELECTRIC POWER DISTRIBUTION ENGINEERING

ANTONIO COLMENAR-SANTOS
ENRIQUE ROSALES-ASENSIO
AND
DAVID BORGE-DIEZ
EDITORS

NOTICE TO THE READER

The Publisher has taken reasonable care in the preparation of this book, but makes no expressed or implied warranty of any kind and assumes no responsibility for any errors or omissions. No liability is assumed for incidental or consequential damages in connection with or arising out of information contained in this book. The Publisher shall not be liable for any special, consequential, or exemplary damages resulting, in whole or in part, from the readers' use of, or reliance upon, this material. Any parts of this book based on government reports are so indicated and copyright is claimed for those parts to the extent applicable to compilations of such works.

Independent verification should be sought for any data, advice or recommendations contained in this book. In addition, no responsibility is assumed by the publisher for any injury and/or damage to persons or property arising from any methods, products, instructions, ideas or otherwise contained in this publication.

This publication is designed to provide accurate and authoritative information with regard to the subject matter covered herein. It is sold with the clear understanding that the Publisher is not engaged in rendering legal or any other professional services. If legal or any other expert assistance is required, the services of a competent person should be sought. FROM A DECLARATION OF PARTICIPANTS JOINTLY ADOPTED BY A COMMITTEE OF THE AMERICAN BAR ASSOCIATION AND A COMMITTEE OF PUBLISHERS.

Additional color graphics may be available in the e-book version of this book.

Library of Congress Cataloging-in-Publication Data

ISBN: 978-1-53614-202-0

Published by Nova Science Publishers, Inc. † New York

CONTENTS

PREFACE

This book constitutes the refereed proceedings of the *2018 International Conference on Renewable Electric Power Distribution Engineering,* which was held on 21st May 2018. *2018 International Conference on Renewable Electric Power Distribution Engineering* intends to provide an international forum for the discussion of the latest high-quality research results in all areas related to *Renewable Electric Power Distribution Engineering.* The editors believe that readers will find following proceedings interesting and useful for their own research work.

This book contains the Proceedings of the *2018 International Conference on Renewable Electric Power Distribution Engineering* held online (https://enriquerosales.wixsite.com/virtualconferences), on 21st May, 2018. It covers significant recent developments in the field of *Renewable Electric Power Distribution Engineering* from an applicable perspective.

ADVISORY BOARD:

Organizing Committee Chair:
Enrique Rosales Asensio, PhD
Departamento de Física, Universidad de La Laguna, La Laguna, Spain
Email: erosalea@ull.edu.es

PROGRAM COMMITTEE CHAIRS:

Enrique González Cabrera, PhD

Departamento de Ingeniería Química y Tecnología Farmacéutica, Universidad de La Laguna, La Laguna, Spain

Email: eglezc@ull.edu.es

Antonio Colmenar Santos, PhD

Departamento de Ingeniería Eléctrica, Electrónica, Control, Telemática y Química Aplicada a la Ingeniería,

Universidad Nacional de Educación a Distancia, Madrid, Spain

Email: acolmenar@ieec.uned.es

David Borge Diez, PhD

Departamento de Ingeniería Eléctrica y de Sistemas y Automática, Escuela Técnica Superior de Ingenieros de Minas de León,

León, Spain

Email: dbord@unileon.es

SCIENTIFIC COMMITTEE:

Clara M. Pérez-Molina, PhD, Universidad Nacional de Educación a Distancia, Madrid, Spain

Francisco Mur-Pérez, PhD, Universidad Nacional de Educación a Distancia, Madrid, Spain

Elio San Cristobal Ruiz, PhD, Universidad Nacional de Educación a Distancia, Madrid, Spain

Pedro Miguel Ortega Cabezas, PSA, Madrid, Spain

Rosario Gil Ortego, PhD, Universidad Nacional de Educación a Distancia, Madrid, Spain

Salvador Ruiz Romero, ENDESA, Barcelona, Spain

Jorge Blanes Peiró, PhD, Universidad de León, León, Spain

May 2018

Editors

In: Renewable Electric Power Distribution ... ISBN: 978-1-53614-202-0
Editors: A. Colmenar-Santos et al.

Chapter 1

THE ROLE OF ENERGY RESERVE AND MAGNETIC ENERGY STORAGE

Enrique-Luis Molina-Ibáñez[1,*],
Jorge-Juan Blanes-Peiró[2] and David Gómez-Camazón[1]
[1]Departamento de Ingeniería Eléctrica, Electrónica, Control, Telemática y Química Aplicada a la Ingeniería, Universidad Nacional de Educación a Distancia (UNED), Madrid, Spain
[2]Departamento de Ingeniería Eléctrica y de Sistemas y Automática, Escuela Técnica Superior de Ingenieros de Minas de LEON, Spain

ABSTRACT

This chapter discusses two essential aspects to take into account for an ESS, that is the regulatory framework and the economic aspect. In particular, it focuses on superconducting magnetic energy storage (SMES) in the Spanish electrical system. An analysis is performed on the legislation and regulations that apply to energy storage systems, which

* Corresponding Author Email: emolina37@gmail.com.

may affect in a direct or indirect manner its inclusion. This is accompanied by an analysis of the legislation in different countries to assess the situation in Spain in this regard, by comparison. Another point to take into consideration, which is crucial for the correct development and inclusion of this type of elements, is the economic viability- showing the costs of manufacturing and maintenance of these systems. Although it is necessary to keep investigating to lower the costs, economic benefits are appreciated, among other things, owing to the increase of the reliability of the electrical network. This increase of the reliability is resultant from a decrease of the cuts of service and the improvement of the quality of the energy.

Keywords: energy storage, superconduction, economic viability, legislation

INTRODUCTION

The growing concern for the environment and climate change over the past years has led to several voices beginning to question the present electric model. For some decades, the use of energy resources of renewable origin [1], which limits the use of polluting sources, has been promoted. Furthermore, the use of strategies that make more rational and efficient consumption possible, such as demand management, has been encouraged.

Considering the inclusion of sources of renewable energy generation in the electrical system, in which the generation of energy by wind turbines and solar photovoltaic panels stands out [2], the use of elements that make energy storage possible is necessary. This is owing to the generation of irregular power that is largely dependent on weather conditions.

Energy storage systems (ESS) can be characterized by different metrics that facilitate the choice of one device or another [3]. The devices that are currently marketed and/or in development are grouped into four major groups: Electrochemistry (different types of batteries), mechanical (FES, PHS, CAES), electrical (SMES, EDLC) and heat.

Approximately 95-98% of the total, storage at the global level is based on PHS owing to the simplicity and maturity of its technology. In spite of this, the quota of ESS compared with that of PHS has grown from less than 1% in 2005 to more than 1.5% in 2010 and 2.5% in 2015 (a growth rate greater than 10%) [4, 5].

These systems should support the proper functioning of the network. It is necessary to bear in mind that the supply and the quality of energy are categorized as a basic need in everyday life. As a result, electricity consumption has been associated with the level of development of a city, region or country, and its evolution has been reflected in its gross domestic product (GDP). Figure 1 shows the variation of the demand for energy in peninsular Spain in comparison with the evolution of the GDP in recent years.

Considering the characteristics of each of energy storage system, there are plenty of cases of the use of elements. The main applications that the ESS are capable of realizing are load tracking applications, energy storage, emergency elements, systems of uninterruptible power supply (UPS), fitness levels of voltage and frequency regulation and elements of protection [7, 8].

The main aim of this chapter is to research about the storage of magnetic energy by using a superconductivity (SMES) system. This type of systems has not reached commercial ripeness for generalized use in a network, as reported [9], owing to different aspects. These problems can be summarised as resulting from high cost of manufacture/maintenance, technical difficulty in the application in different environments and the lack of normative support.

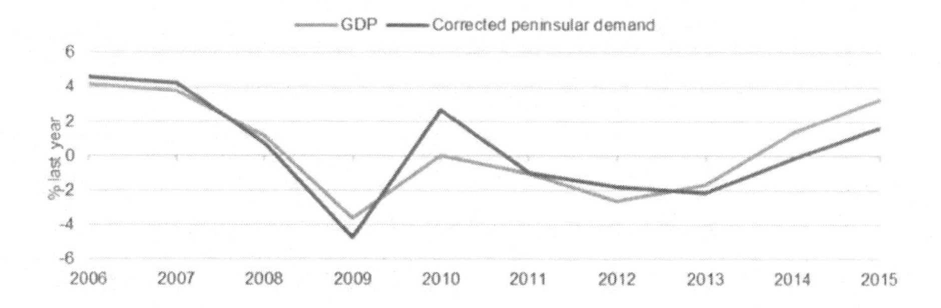

Figure 1. Comparative GDP vs Energy Demand [6].

An SMES system allows the storage of energy under a magnetic field because the current through a coil is cooled at temperature below the critical temperature of superconductivity. The system is based on a superconducting coil, a cooling system that allows the critical temperature to be obtained, and an electrical and control system for the adaptation of currents and the optimization of the process.

Given the large spectrum of research concerning the solution of the problematic technique for the inclusion of SMES systems in different configurations, this article focuses on two important aspects to enhance its use in power system, that is, legislative and regulatory aspects and the economic aspect.

To perform a correct analysis of this type, the status of capacity of the main characteristics of this type of ESS must be born in mind, as summarised in Table 1. The characteristics of these systems may vary depending on the type of SMES. SMES are categorized according to their critical temperature (Tc), LTS (NbTi) and HTS (YBCO, BSCCO), and according to the configuration for their use [10-17], in which the optimization of the performance of the device is searched for in different processes and systems. This implies betting for the investigations of new alloys with higher critical temperature than the HTS [18], the optimization of the elements of electrical adaptation, as well as investigations in the systems of regulation and control [19] or the study of the inclusion of these systems in the microgrids/smart grids [20, 21].

Table 1. Main characteristics of a SMES [3, 7, 8, 22-38]

Daily self-discharge (%)	Energy Density (Wh/L)	Specific energy (Wh/kg)	Power Density (W/L)	Specific power (W/kg)	Power (MW)	Response time	Discharge time	Suitable storage duration	Efficiency (%)	Lifetime (yr)	Lifetime (cycles)
10–15	0.2-6	0.5-5	1000-4000	500-2000	0,01-10	<10ms	ms-min	min-h	>90	20+	$5\cdot10^4$

Owing to the characteristics of these type of systems, applications are restricted to a group of potential uses focused on electrical power systems, which are essential for providing an adequate quality system. Table 2 shows the applications of this type of ESS.

The methods used to carry out the investigation of this article are outlined in Section 2. In this section, the legislation on ESS for the application in the Spanish electrical system is shown as an example of a system in which the penetration of renewable energies has had a high impact. The main problem that prevents the complete maturation of the system, the economic casuistry, and a feasibility analysis of such a system are also addressed in this section. This is why the economic impact of its use in the electrical system, from manufacturing costs to maintenance costs, is analysed. The results of the economic study concerning the inclusion of SMES storage systems in the electricity network are presented in Section 3. This allows the possible economic benefits of the inclusion of these systems in the electricity network, and other indirect benefits to be determined.

The legislative and normative issues are discussed in Section 4, both in terms of standardization of the equipment and regulation, conditioning the implementation of SMES systems and its competitiveness with other systems [42]. Finally, Section 5 is reserved to show the main conclusions obtained from the normative and economic study of these systems.

Table 2. Applications of SMES [7, 8, 29, 38-41]

Application area	Standing reserve	Emergency and telecommunications back-up power	Load following	Uninterruptible Power Supply (UPS)	Voltage regulation and control
	NO	NO	YES	YES	YES
	Black-start	Frequency regulation	Integration of renewable power generation	Grid fluctuation suppression	Spinning reserve
	YES	YES	YES	YES	YES

Material and Methods

For this case study, an analysis differentiated in two parts has been realized. On the one hand, the Department of Energy of Spain has the legislative and normative information relative to the whole process of generation and energy consumption. All legislation approved in relation to the Spanish electricity system is published in the BOE (Official Bulletin of the State), this being an essential reference. This legislation affects, in a direct or indirect way, the systems of energy storage. With regard to the legislation in other countries, information can also be found primarily in the concerned ministries or departments of the State. The normalization and standardization are detailed in Appendix A.

Various documents were analysed for the economic study: the economic cost of the construction of SMES, the potential economic benefits of the inclusion of SMES in the electrical system and the environmental benefit use of an ESS.

Finally, the amount of harmful gasses generated from coal consumption was analysed and the possible saving from the inclusion of the ESS. For the quantity of generated gasses it is necessary to bear in mind the type of coal that is mainly consumed and the proportion of gasses generated by typology for each kilogram of consumed coal. With this information, it is possible to perform an analysis of the large amounts of these gasses that might be avoided thanks to the ESS, as well as determine the economic implications of reducing the emission of these gasses.

Theoretical Framework

At the legislative level, in Spain there is no law or specific regulations that enable the research, development and implementation of these systems. However, the inclusion of other ESS as kinetic energy storage has been promoted. A laboratory prototype has been developed which an emulator for railway catenary, an emulator of consumption of electric vehicles and a unit for the storage of energy based on ultracapacitor have been integrated and tested on a system installed in the underground of Madrid [43]. Also a flywheel of 25 kW, 10 MJ has been adapted for operation in a microgrid, for the application as compensation during consumption peaks and regulation of frequency [44].

In the case of the Spanish electricity system, we should take into account the different policy levels, in order to ensure an adequate inclusion of SMES systems, enhancing its use and regulation in manufacturing systems. These levels can be summarised as:

- European Union (EU), through the corresponding Regulations or Directives [45].
- National, through ordinary laws, Royal Decree Law or Regulations (Royal Decree, Ministerial Order, Circulars, Resolutions, etc.) [46, 47].
- Other regulations of regional application, such as Decrees or Orders.

The legislation relating to the regional level is very limited in regard to the inclusion of ESS of large or medium scale. Despite this, Spain may grant economic aid to encourage the installation on a small scale, for micro-SMES systems of local storage.

tCalculations

There are several studies that seek to perform an economic analysis on the ESS in a general way [16, 48-54]. In this way, the costs can be grouped in Invested Capital (C_I), Capital of Operation and Maintenance ($C_{O\&M}$) and Financial Capital (C_F), or Capital of Investment.

In spite of everything, it remains that the total storage is:

$$TSC(\$) = C_I(\$) + C_{O\&M}(\$) + C_F(\$) \tag{1}$$

In which the total invested cost, C_I, can be defined as the sum of costs of material, construction and commissioning, own of this ESS. For this analysis of costs, it is necessary to carry out a revision of the main components listed previously. These systems are mainly composed of:

- Superconductive coil
- Criogenization system
- Electrical system
- Monitoring and control system

The adequacy of analysis takes into account materials and configuration to be treated, as the cost of the superconductor element itself, which is the most expensive element of the device, in either LTS or HTS devices. Figure 2 shows an example of a coil and the main elements of the SMES storage system.

The investment costs can be grouped into three subgroups:

$$C_I(\$) = C_{st}(\$) + C_e(\$) + C_{BOP}(\$) \tag{2}$$

In which:

C_{st} ($) is the cost of construction of the storage system,

C_e ($) is the cost of the electrical system of the device, and

C_{BOP} ($) is the cost of balance of the plant and cost of the auxiliary system.

Despite how meticulous this analysis can be, in which you can compute the minimum cost of the most basic element, it possible to be simplified using the sizing of the device, that is:

$$C\_st\ (\$) = (C_E \cdot E)/\eta \tag{3}$$

$$C_e(\$) = C_P \cdot P \tag{4}$$

$$C_{BOP}(\$) = C_{BOP}(\$/kW) \cdot P \text{ or } C_{BOP}(\$) = C_{BOP}(\$/kWh) \cdot E \tag{5}$$

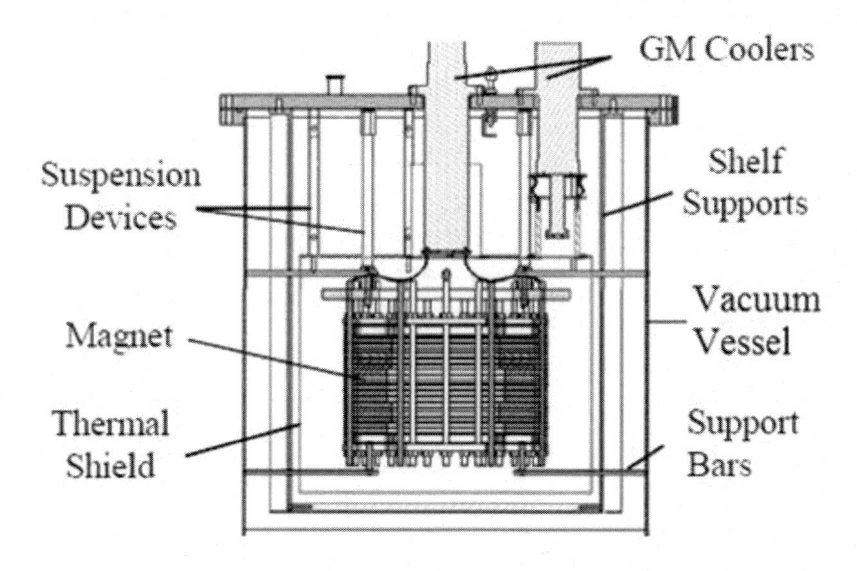

Figure 2. SMES System [55].

In which:

C_E is the energy cost ($/kWh),

E is the stored energy (kWh),

η is the efficiency of the system,

C_P is the cost of power ($/kW), and

P is the capacity of power (kW).

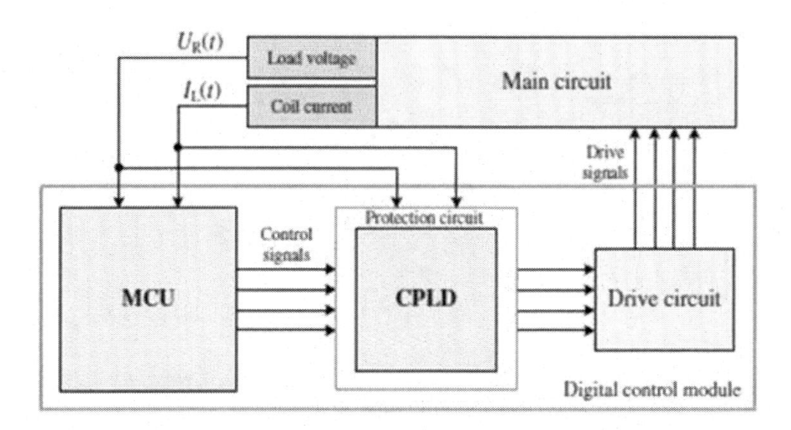

Figure 3. Control module of a SMES system [56].

In equation (5) it is possible to use on formula or another depending on the available data for the analysis.

The cost of balance of the plant incorporates the control module that enables the proper functioning and performance of the system. Figure 3 shows a schematic diagram of a control module but it can vary depending on the configuration blocks (D-SMES), its application or if it is part of some type of hybrid storage system.

The wear of the materials in the working conditions, electrical or thermal, must be considered in the costs of maintenance and operation. It is also important to take into account the energy expenditure at the criogenization to maintain the temperature at the optimum operating conditions, a variable expense that can be supplanted by annex systems. It is estimated that a typical cooling system requires approximately 1.5 kW per MWh of stored energy [57].

Furthermore, the skilled labour needed for the operation of the system operation should be borne in mind. As with other factors, these operating costs are variable and can be approximated as a function of the capacity of power and the years of operation.

$$C_{O\&M}(\$) = C_{O\&M}(\$/kW) \cdot P \cdot k \quad (6)$$

Finally, we find a variable term, depending on the interests of the investment and the years. Normally this cost is characterized by:

$$C_F(\$) = C_I(\$) \cdot \delta \quad (7)$$

With a multiplier factor δ which is given by:

$$\delta = (r \cdot (1+r)^k)/((1+r)^k - 1) \quad (8)$$

In which:

r is the interest of the investment, and

K is the time of life, in years.

After analysing the costs of the manufacture and maintenance of the SMES systems, the economic advantages of the use of these systems must be analysed. To do this, the information of the availability is obtained in the Spanish electrical system.

Energy not supplied (ENS) measures the power cut to the system (MWh) throughout the year resulting only from network service interruptions. Only interruptions of over a minute duration zeros of tension are counted. In this case, the inclusion of an SMES system would reduce the cuts that are limited duration, owing to its low energy density. For electricity cuts of longer duration, hybrid systems could be implemented [58]. Another solution could be the improvement of the energy density of these systems; an extensive number of studies have been performed on this topics [55, 58, 59-61].

Average interruption time (AIT) is defined as the relationship between the energy not supplied and the average power of the system, expressed in minutes:

$$TIM = HA \cdot 60 \cdot ENS/DA$$

In which:
HA is the hours per year, and
DA is the annual demand of the system in MWh.

Appendix A shows some of the aspects to keep in mind about regulation and economic facets not indicated previously but which may have importance for the compression of some aspects.

RESULTS

To evaluate the cost of the storage of the SMES system and determine its economic viability, it is necessary to bear in mind that different characteristics play an important role in the manufacture of these elements, such as the size of the element of storage.

This study focuses on systems destined for the regulation and storage of the Network of Transport and Distribution, so neither systems Micro-SMES nor Mini-SMES would be described; their storage capacity is more limited and they would be destined for domestic use.

Economic Analysis

The costs of an ESS tend to be according to the capacity of potency and/or energy, that is, $/kW or $/kWh. In recent years the processes for the production of SMES modules as well as the auxiliary systems have been improved, the price of the manufacture of elements have been lowered, in some cases replacing them with elements that have the same properties but are more accessible economically. All this has allowed a variety of costs across a wide range, as shown in Table 3.

The price of a HTS in recent years has been approximately 35 $/A·m for a BSCCO and 15 $/A·m for a YBCO, and it continues to decrease [56]. This also happens with other ESS, for which it is estimated that the costs will be reduced by approximately 20% on average, as shown in Figure 4 for other technologies.

As example, using the information of the text of S. Sundararagavan [52], Table 4 shows the costs, which depend on the characteristics and on the materials.

With these data, and considering the study by Ren et al. [30] in which there is a SMES system Energy/Power (MWh/MW) = 6,49/1,52, as well as an interest of r = 10%, the entire cost of the project is:

Table 3. Price range of an SMES system [7, 22, 24-29, 31, 36-38, 59-62]

	$C_E(\$/kWh)$	$C_P(\$/kW)$
SMES System	700-10.000	130-515

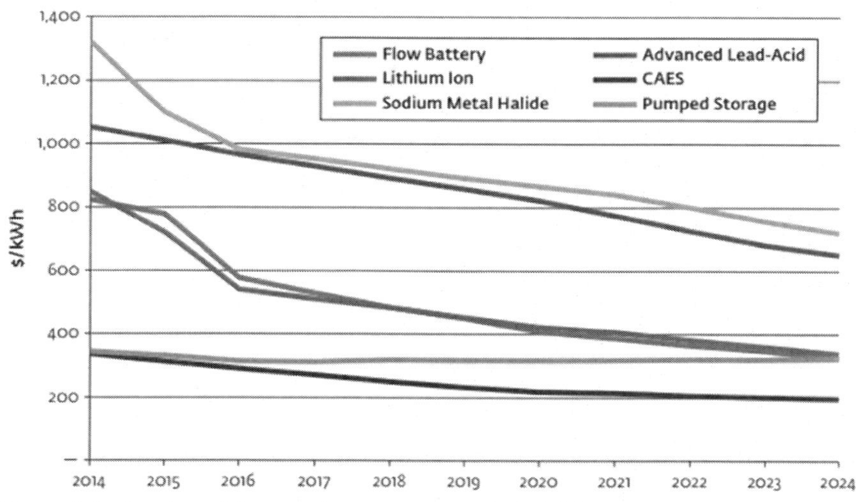

Figure 4. Estimation of the cost for storage technology [63].

Table 4. Example of costs of a SMES system [52]

Technology	Energy cost ($/kWh)	Power cost ($/kW)	Balance of plant cost ($/kWh)	Operation & maintenance cost ($/kW)	Efficiency (%)	Lifetime (yr)
SMES	10	300	1.5	10	95	20

C_I ($)	$ 68.781.524,47
C_{OM} ($)	$ 304.000,00
C_F ($)	$ 8.079.052,06
TSC ($)	$ 77.164.576,53

In this study Ren et al. show a cost of approximately 1.358.300 $/year, with an average useful life of more than 20 years, for a total of 27.166.000$. These data indicate the wide ranges in the projects of installation of a system of this type, influenced by different factors and technologies.

With the obtained data, a comparison could be performed show the impact of this cost on the budget of a Spanish city of importance, such as

Zaragoza, which has a budget of 744,3 M€ [64] (808 M$), so the creation and operation of such a system would account for approximately 7.7% of its overall budget.

Economic Benefits

The information of the availability and quality of electricity supply provided by the system operator in the Spanish electrical system (REE) must be analysed to obtain the possible economic benefits. This information for the electricity transport network from 2011 is given in tables 5, 6 and 7 [65].

From this, the total direct losses from energy that has been generated but not supplied can be obtained, as shown in Figure 5. This figure is generated with data from REE.

Table 5. Peninsular transport network

Peninsular transport network	2011	2012	2013	2014	2015
Network availability (%)	97.72	97.78	98.2	98.2	97.93
Energy not supplied (ENS) MWh	259	113	1.126	204	52
Average Interruption Time (AIT) min.	0.535	0.238	2.403	0.441	0.111

Table 6. Balear transport network

Balear transport network	2011	2012	2013	2014	2015
Network availability (%)	98.21	98.07	97.96	98	96.87
Energy not supplied (ENS) MWh	35	7	80	13	7
Average Interruption Time (AIT) min.	3.194	0.678	7.366	1.205	0.642

Table 7. Canarian transport network

Canarian transport network	2011	2012	2013	2014	2015
Network availability (%)	98.95	98.91	98.3	98.37	96.76
Energy not supplied (ENS) MWh	17	10	3	64	29
Average Interruption Time (AIT) min.	1.023	0.613	0.177	3.938	1.763

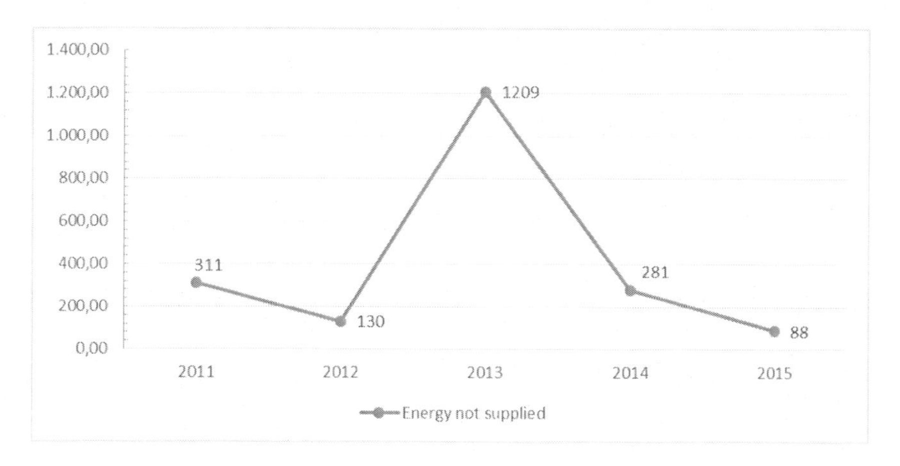

Figure 5. Losses owing to cuts of service [65].

It is necessary to add the indemnifications of the electrical companies to the users to the losses generated by the cost of generation. The minimum established quality according to the regulation will bear in mind both the number of cuts and the total amount of time, in a year, in which there has been no supply, according to the area and how it is categorized.

A user is entitled to receive a discount on the bill for the first quarter of the year after the incident. The clients may also request another type of compensation in case any of their goods are damaged owing to power cut.

The National Commission of the Markets and the Competition (CNMC) has valued penalties to the Spanish electrical distributors at 52,5 M€ for their network losses in 2016 [66].

Furthermore, it is necessary to count the economic losses produced by the time of non- operation of different factories and different productions. In this case, it is more complicated to know the exact amount of the losses, because it depends on factors such as the type of industry, the time it occurs, or the location. It is at this point at which the most significant losses occur.

The industries in which a continuous process is important, in which the shutdown of the production can result in a high amount losses, because a determined time is needed to restart engines. It is in this case that the SMES systems have an important role; the starter time would be reduced considerably owing to the high thickness of potency.

Environmental Benefits

In addition to the direct economic benefits, there are also indirect benefits, which include the environmental benefits. These environmental benefits allow a reduction of energy produced by sources of pollution, such as coal. The consumption of different types of coal produces substances that are harmful to human beings and can produce alterations in the biological cycles of the species, as well as other consequences. These consequences may involve an increase in the costs of treatment of diseases, treatments for environmental recovery as well as treatment for the protection of architectural elements produced as a consequence of the increase in the proportion of different substances diluted in the air.

A great variety of harmful substances appears with the consumption of coal because of its composition. This is the reason why it is necessary to perform an analysis of the amount of derived but not consumed coal from the use of elements of energy storage. The quantity of not consumed coal (CNC) can be estimated as a result of the use of the ESS with the following formula:

$$CNC = E_{SESS} \cdot h_{\%C} \cdot R_{conv}$$

In which:

E_{SESS} is the energy provided by ESS (kWh),

$h_{\%C}$ is the percentage of energy provided by sources of coal (%), and

R_{conv} is the conversion factor of energy of the coal ((kg(Coal))/MWh).

The variation of the energy mix during the day must be taken into account, so the formula changes to:

$$CNC_D = \left(\sum_{j=0}^{23} E_{SESS_j} \cdot h_{\%C_j} \right) \cdot R_{conv}$$

This formula considers the factor of energy conversion of coal constant, but depending on the mix of used coal it may vary.

Table 8. Emission factor of the main substances [67]

	Emission factor (χ)	Units
Carbon dioxide. CO_2	2.29700	Kg of CO_2/kg of coal
Carbon monoxide. CO	0.00025	Kg of CO/kg of coal
Sulfur anhydride. SO_2	0.05510	Kg of SO_2/kg of coal
Ammonia. NH_3	0.00086	Kg of NH_3/kg of coal
Nitrogen dioxide. NO_X	0.01100	Kg of NO_2/kg of coal

With this, the amount of substances emitted to the atmosphere can be calculated. This depends on the emission factor of the different substances. Table 8 shows the emission factor of the main substances:

As a result, it is possible to obtain the quantity of substances released from the coal that are not released owing to the use of ESS by using this formula.

$$R_x = \chi_y \cdot CNC$$

In which:

y; It can be: CO_2, CO, SO_2, NH_3, NO_X

The information from the last few years in Spain of the coal consumption is summed up in Table 9.

Table 9. Coal statistics in Spain [67, 68]

	Average annual generation (%)	Energy generation (GWh)
2009	12.50%	34.793,03
2010	8.30%	23.700,61
2011	15.60%	43.266,69
2012	19.20%	53.813,42
2013	14.70%	39.527,56
2014	16.50%	43.320,30
2015	19.90%	52.789,04
2016	14.50%	37.474,06

Table 10. The amount of substances generated by the consumption of coal for the generation of electricity [68]

	Amount of substance generated per year [ton]				
	CO_2	CO	SO_2	NH_3	NO_X
2009	14.027.869,55	1.526,76	336.497,87	5.252,05	67.177,43
2010	9.555.625,39	1.040,01	229.218,53	3.577,64	45.760,50
2011	17.444.287,85	1.898,59	418.450,27	6.531,17	83.538,17
2012	21.696.521,14	2.361,40	520.452,03	8.123,21	103.901,49
2013	15.936.743,31	1.734,52	382.287,57	5.966,74	76.318,75

The amount of harmful substances is obtained from this information. These substances are generated by coal consumption for the generation of electric power, during the year, in tons, as shown in Table 10.

For these reasons, this is one of the goal for using this systems for the storage of electric power. It is necessary to bear in mind that this information only corresponds to the generation of substances derived by coal consumption. It would be necessary to add the use of other sources for the generation of electricity, such as those of a combined cycle system or fuel oil.

From these data, it is possible to estimate the amount of coal saved as a result of using energy storage systems. Knowing the percentage of energy supplied by coal sources, the energy supplied by the energy storage sources and the energy conversion factor of the coal [71], the carbon saved and the CO_2 emission not made as a result of saving coal were calculated and are shown in Table 11.

Table 11. Saved tons of carbon and CO_2 by ESS [71]

	% Carbon	E_{SESS} (MWh)	CNC	CO_2
2010	8,30%	4.457.782,58	3.909,58	8.980,30
2011	15,60%	3.214.959,82	5.299,48	12.172,90
2012	19,20%	5.022.547,79	10.189,62	23.405,56
2013	14,70%	5.957.844,99	9.254,21	21.256,92
2014	16,50%	5.329.590,05	9.292,03	21.343,79
2015	19,90%	4.520.094,18	9.504,59	21.832,04
2016	14,50%	4.819.413,08	7.384,05	16.961,17

These data are obtained thanks to the energy produced by the PHS systems, because they are the main storage system in Spain. The energy obtained by the other systems can be considered residual at the moment.

Discussion

In the current stage in which high capacity SMES systems are (research/pre-sale), economic and financing support and a legislation that regulates their application are important. Therefore, adequate regulation at different levels would allow this storage system to be developed and to provide its advantages or, conversely, to be discarded for inclusion in an electrical system in which the use of other systems is more technically or economically appropriate.

The potential storage of energy that the Spanish electrical system has and the predisposition for the inclusion of the ESS are notorious, as shown in Figure 6. The power of installed storage and developed storage projects are represented in this figure.

Community Legislation (EU)

There are numerous resolutions of the European Parliament that aim to promote the use of renewable energy and the reduction of GHG emissions. For example, obligatory targets for 2020 [72], the resolution of February 2014 [73] for the Horizon of 2030 or the Roadmap of the Energy for 2050 [74], among others [75].

Furthermore, there are resolutions of the European Parliament which demand the creation of a long-term system of common incentives to scale the EU in favour of renewable energy sources [76]. These resolutions also support the technologies of smart grids [77], as well as the microgeneration of electricity and heat at a small scale [78], which seeks to support the personal energy consumption of citizens, as well as the need to establish incentives that encourage the generation of energy at a small scale.

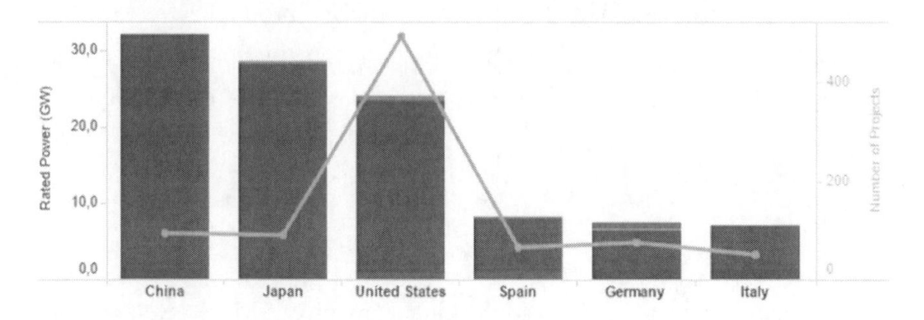

Figure 6. Stored power – Storage projects [5].

To realize a transition to an energy model such as the one proposed by the Parliament in Europe, it is necessary to provide flexibility to the European energy system through the improvement of the technologies of storage of energy.

Innovation activities relating to storage at the local level as, for example, in residential areas or industrial estates, seek to create synergies between technologies and to improve connections of a secure and stable form, even in remote areas without a sufficient connection to the electrical network.

For the large-scale storage, the investment seeks to ensure high rates of penetration of renewable energy sources to cover high electricity demands for longer periods of time. Furthermore, the innovative actions must ensure the integration and management of networks and synergies between an electric network and others.

It also gives importance to the development and improvement of the technologies of energy storage that achieve better results with lower costs. For each technology, the profitability cost-benefit is being studied and analysed using scenarios and simulations, the expansion of the electricity network, the incorporation of other storage systems and the management of the energy economy.

One of the examples of this type is the project “Grid+ Storage” [79]. It identifies actions focused on the integration of the energy storage in the distribution networks with the target of making them more flexible.

Concerning the main regulation relative to the ESS, the European legislation that appears in Table 12 must be taken into account.

Table 12. Main European legislation

Norm	Date	Ambit	Summary
The Treaty on European Union and the Treaty on the Functioning of the European UNION [77]	2010	Charter of Fundamental Rights of the European Union.	- To guarantee the functioning of the market of the energy. - To guarantee the safety of the energy supply in the Union. - To encourage the energy efficiency and the energy saving as well as the development of new and renewable energies. - To encourage the interconnection of the energy networks
Directive 2009/28/EC of the European Parliament And of the Council [78]	23 April, 2009	Concerning the promotion of the use of energy from renewable sources and which modify and repeal the Directives 2001/77/CE and 2003/30/CE	- Supports the integration into the network of transport and distribution of energy from renewable sources and the use of systems of energy storage for the variable integrated production of energy from renewable sources
Directive 2009/72/CE of the European Parliament and of the Council [79]	13 July, 2009	Concerning common rules for the internal electricity market.	- Establishes common rules for the generation, transport, distribution and supply of electricity, as well as rules concerning the protection of consumers, with a view to improve and integrate competitive markets of the electricity in the EU.
Directive 2012/27/EU of the European Parliament and of the Council [80]	25 October, 2012	Concerning the energy efficiency, which modify the Directives 2009/125/CE and 2010/30/UE, and which repeal the Directives 2004/8/CE and 2006/32/CE	- Shows the different criteria of energy efficiency for the regulation of the network of energy and for the tariffs of the electrical network
Regulation (EU) No 347/2013 of the European Parliament and of the Council [81]	17 April, 2013	Concerning the guidelines for trans-European energy infrastructures.	- The projects related to transport and storage of energy should promote the use of renewable sources, storage systems, guaranteeing the supply, opting for financial aid from the Union in the form of grants.

National Legislation

The European directives involve a series of laws to the Member States such as Spain. These laws are listed in Appendix B. This appendix shows the two main laws governing the electricity sector in Spain, Law 54/1997 [85] and Law 24/2013 [86]. These laws have made possible the liberalization of the electrical sector in Spain. One of the points that distinguishes Law 24/2013 from the previous one is the disappearance of the previous "special regime", which included renewable energies, cogeneration and waste. Article 23 of this law indicates that electric energy producers make economic offers of energy sales in the daily market, with the particularity that all production units must make offers to the market, including those of the former special regime [86].

In these law, as in the others listed in Appendix, SMES storage systems are not refereed to explicitly but the features and functions of the different components of an electrical system are discussed. That is why these and other regulation on the table are important in relation to the SMES storage system and its applications.

Table 13. Operative Procedures

Operative Procedures	Ambit
P.O. 1.2 [87]	Allowable levels of load network.
P.O. 2.1 [88]	Demand forecasting.
P.O. 2.5 [88]	Maintenance of units of production plans.
P.O. 3.1 [89]	Programming of the generation.
P.O. 3.7 [89]	Application of limitations to deliveries of energy production in non-resolvable situations with the application of the adjustment of the system service.
P.O. 3.10 [90]	Resolution of restrictions by assurance of supply.
P.O. 7.4 [91]	Complementary service of voltage control of the transport network.
P.O. 8.2 [92]	Operation of the system of production and transport.
P.O. 13 [93]	Criteria of the planning of the networks of transport of the insular and extrapeninsular electrical system.
P.O. 13.1 (94)	Criteria of development of the transport network.
P.O. 13.3 [95]	Transport network facilities: criteria of design, minimum requirements and verification of their equipment and commissioning.
P.O. 15.2 [96]	Management service of demand of interruptibility service.

Table 13 indicates the Operative Procedures (OP, Appendix A) that can affect the ESS and that are specifically named in the regulations owing to their application in an electrical system.

The Operative Procedures seek the technical adequacy of the elements in the transport network. As for storage systems, these procedures focus on pumped storage systems. The companies that own the plants have the obligation to transmit different data to the system operator, such as quotas and volumes stored in the reservoirs or foreseeable variations of availability of the pumping groups, on a weekly basis [87].

It should be borne in mind that storage systems can be considered production units at any given time, so they must meet the requirements of the system operator [92] as well as ensure supply [90] and interruptibility [96].

The main functions of the system operator are presented in the OPs, such as generation scheduling, solution of technical restrictions, resolution of generation-consumption deviations or complementary service of tension control of the transport network, in which it can play the essential role of ESS [92].

It is possible to observe the varied legislation that can affect the ESS as elements of the electrical system. This legislation largely focuses on the part of generation and transportation of energy from the electrical system, with consideration of the system operator (REE). They are based on the technical and regulatory aspects that allow the involvement of the State and society through public subsidies for its development improvement. The importance of knowing the legislative structure and the context regulatory in the electrical system lies here, to encourage the inclusion of these elements, both in the transport network and in the distribution, and to be able to make a synthesis of these aspects that may directly or indirectly affect the inclusion of the SMES storage systems.

The management of subsidies and incentives in the implementation of renewable energies, (and consequently of the storage systems) is the main focus of action, as well as the regulation of technical aspect for its proper connection to the network.

Regulation and Standardization

Appendix C shows the standard UNE that is applied to manufacturing processes, research and development as well as to the operation and maintenance of these SMES systems. It must be born in mind that these systems can also affect standards as the protections of wiring, electrical protection systems, and a long list which focuses on the storage system itself. Much of this regulation will depend on the characteristics, size and application of the system to apply. For this reason, it is necessary to take into account the elements of construction and the type of device to be able to apply this type of standardization.

Comparison with Other Countries

In addition to have in mind the grade of adaptation of the ESS in the electrical systems, it is necessary to take into account that the electrical networks are interconnected and that the way of operation of one can affect others. This shows the importance of considering the regulation level of other countries to see the implication of the regulation in the inclusion of these ESS.

Furthermore, the need to know the regulations of other countries with a similar development, and referents in that field, makes it possible for these regulations, or part of them, to be adapted to the Spanish electricity system with the necessary changes with the security of its correct operation.

Therefore, the electrical regulation field of some countries was revised. USA, Japan and Germany can be highlighted for the creation and implementation of ESS of the type SMES, with different characteristics and situations. The made devices they can stand out are:

- Chubu Electric Power Company (Japan): Material Bi-2212, Energy 1 MJ [59].
- Los Alamos Laboratory (USA): Material NbTi, Energy 30 MJ [60].
- ACCEL Instruments GmbH (Germany): Material Bi-2223, Energy 150 kJ [61].

Table 14. Comparative table USA-Japan-Germany [97-100]

	USA	Japan	Germany
Main Energy Law at National level.	Energy Policy Act of 2005 PL 109-58	Basic Law of Energy Policy - 4th Strategic Plan of Energy (enerugi kihon keikaku)	Erneuerbare-Energie-Gesetz 2017
Renewable Objectives	Does not specify	3rd Plan: 50% (2030) 4th Plan: Does not specify	45% (2025)
Financing of Renewable Energies	The law provides loans guarantees to the entities that develop or use innovative technologies that prevent the sub-production from greenhouse gases.	Sets that renewable energies will expand their market rate by 10% thanks to the "Feed-in tariff" (FIT). FIT is a remuneration set by the government for energy injected into the network.	Sets the FIT as a mechanism of incentives for renewable energy. The cost of the FIT moves to the users through the finalist EEG rate.
Research and Development	It encourages the research and the development of new elements of generation and energy efficiency that make possible the decline of GHG	Increase in financing of renewable energy and energy efficiency projects. Japan is one of the largest exporters of technology in the energy sector and has a strong program of research, development and innovation backed by the Government.	Aid for the new projects related with the renewable energies and the facilities that are considered for domestic use or that do not come into the consideration of intensive exploitation.
Other.	In Sec. 925 it is explicitly indicated that there should be a focus on storage systems and systems of high-temperature superconductivity research	The storage system introduction is promoted in an explicit way using batteries to ensure the supply and quality. It also refers to other system of electrical energy storage, as the PHS or fuel cells.	Electricity used for temporary storage operators of transport networks to the payment of the surcharge EEG shall not apply if the power is removed from the installation of electricity storage only for feedback on the electricity in the network system.
Example SMES	Los Alamos National Laboratory: Material NbTi, Energy 30 MJ	Chubu Electric Power Company: Material Bi-2212, Energy 1 MJ	ACCEL Instruments GmbH: Material Bi-2223, Energy 150 kJ

Table 14 shows the comparison of these three energy models with the action plan and the main standard. The table focuses on measures to take into account on the basis of renewable energies and their promotion at the institutional level. It is explained in more detail in Appendix D.

Apart from these examples, the Paris Conference on Climate [101] is also important. It was celebrated in December 2015, during which 195 countries signed the first binding agreement on global climate. One of the most important points was to ensure that the global average temperature rise was kept below 2°C above pre-industrial levels. The renewable systems will play a key role in achieving this target and all elements influence.

CONCLUSION AND POLITICAL IMPLICATIONS

Considering the importance and the impulse of the generation of energy through renewable sources in the energy mix, the elements that orbit around it become vital for the correct inclusion of renewable sources without an impact on the supply quality.

The need to know the regulation that affects the storage systems, directly or indirectly, implies realizing the potential inclusion of these elements. There are a few legislations in Spain with direct implications for storage systems but there are regulations that indirectly affect them, despite the fact that the contributions from institutions in this regard have been reduced in recent years. Not having a specific legislation can negatively affect SMES systems in favour of other more mature systems, such as batteries or PHS (despite the geographical limitations of these).

The rise of renewable energy at the expense of other less clean energy has enabled the development and investment, both public and private, in storage systems. These initial investments and specific regulation are indispensable to allow the competitiveness of very advantageous elements but in an unfavourable commercial position.

Another critical lever on the inclusion of any element is the economic vision of a project. The technological complexity derives from the materials and the cooling system, which involves always maintaining the coil at a temperature below the critical temperature of the material of the coil. This complexity involves some manufacturing and maintenance costs of SMES systems that make it difficult to apply in the transport network of the electricity network in Spain.

Therefore, the applicable legislation to the storage systems and the economic viability of its construction, commissioning and maintenance, as well as the interrelation between both can be determinants for eventual insertion into the electrical network. The solution seems obvious: greater institutional involvement in the development and research of storage systems and their components, which make possible the improvement of the technical capabilities of the systems at a lower cost. This involvement can not only come from grants from public institutions, but also through tax aid, shared financing or other appropriate formulas that enable this development.

It is a fact that the inclusion of renewable sources of energy and the ESS as a result of its intermittent and unstable characteristics, can bring great benefits of different types: social, environmental and economical. It is necessary to invest in the development of SMES systems, or hybrid systems that combine the strengths of high energy density of the batteries with the high power density of SMES systems.

APPENDICES

Appendix A.

A.1. Normative Aspects

All community legislation and regulation must be translated in regulatory laws in every Member State. This makes possible the adequacy of the activity to the proposed one of the European regulation. The EU has two bodies with the power to adopt binding decisions and to solve the problems that the national regulatory authorities are unable to resolve:

- The Agency for the Cooperation of the Energy Regulators (ACER).

- The European Network of the Operators of the Systems of Transmission of Electricity

Furthermore, it is necessary to bear in mind that SMES storage systems are in the part of the transport and distribution of electrical system. It is work of the company dedicated exclusively to the transport in the Spanish electrical system, Electrical Network of Spain (REE). This company acts as the system operator and has some technical and instrumental protocols, called Operative Procedures (OP). An adequate technical management of electrical system peninsular and electrical systems outside the Iberian Peninsula is guaranteed. These OP are approved by resolutions of the Ministry of Industry which seek to guarantee the stipulation in the Law.

The study and development of the standards is the responsibility of a number of institutions that have the legal power to its realization. The ISO (International Organization for Standardisation) [102], is in charge of the ISO standards. It is formed by 163 agencies of normalization of their respective countries.

At the European level are the European Committee of Standardization (CEN) [103] and the European Committee for Electrotechnical Standardisation (CENELEC) [104], which are responsible for the development of the European Norms (EN).

The Spanish case focuses on the regulations created by the Spanish Association for Standardisation and Certification (AENOR) [105], which disseminates the Spanish rules that are identified with the acronym UNE (a Spanish Norm). AENOR is the Spanish representation in the international standardization organizations ISO and IEC, European CEN and CENELEC, and the Pan American Commission for Technical Standards (COPANT) [106].

To take into account the specific normative in the manufacture and inclusion of the SMES systems, its construction schema must be considered. A possible schema of a SMES storage system, either LTS or HTS, is shown in the Figure A.1.

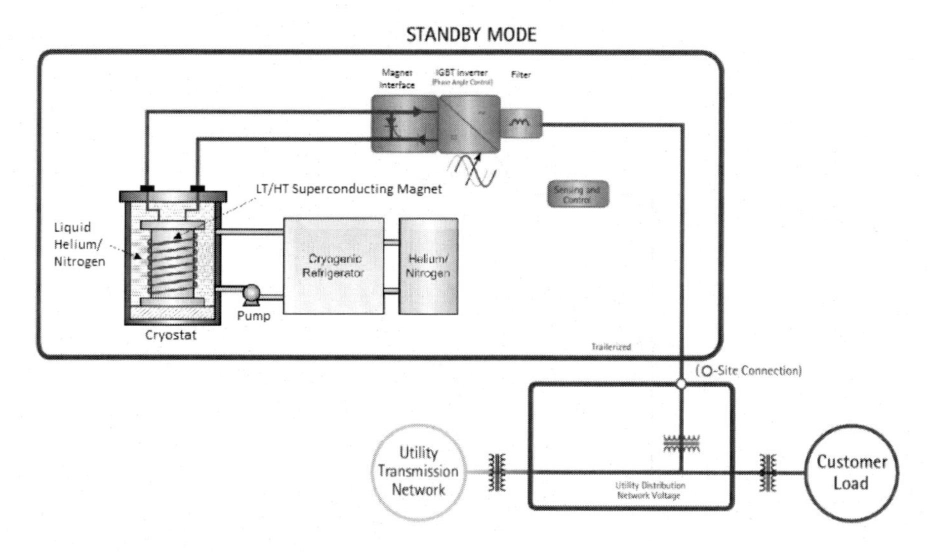

Figure A.1. Basic scheme of a SMES system [107].

A.2. Economics Aspects

In this sense, it is necessary to emphasize that the first used SMES for experimentation and for commercial use was designed by Los Alamos National Laboratory (LANL) and constructed for Bonnevile Power Company in 1982. It was in use for 5 years and was dismantled for investigation [60, 108].

This project had an energy capacity of 30 MJ and it was used to stabilize the potency system, because it cushioned the oscillations in a line of transmission of 1500 km long. In this case, the cost of construction of this system of storage was distributed in the following way:

- Superconductive coil, 45%.
- Structure, 30%.
- Labor, 12%.
- Converter, 8%.
- Cooling system, 5%.

Appendix B.

Table B.1 shows a list of legislation related to the Spanish electricity system and which affects, directly or indirectly, the implementation, use and development of storage systems.

Table B.1. Main Spanish legislation relative to the electrical system

Norm	Date	Ambit
LEY 54/1997 [85]	27 November 1997	Basic Law of the Spanish electricity sector
REAL DECRETO 2019/1997 [109]	26 December 1997	It organises and regulates the electricity production market
Real Decreto 1955/2000 [110]	1 December 2000	Regulates the activities of transport, distribution, marketing, supply and installations of electricity authorisation procedures.
Real Decreto-Ley 6/2009 [111]	30 April2009	Certain measurements are adopted in the energy sector and the social bond is approved.
Real Decreto 134/2010 [112]	12 February 2010	The procedure of resolution of restrictions by supply guarantee is established and the Royal decree 2019/1997, of December 26 which organizes and regulates the market of production of electric power, is modified.
Real Decreto -Ley 6/2010 [113]	9 April 2010	The content of articles 1, 9, 11 and 14 of law 54/1997 of 27 November are modified, in the Electricity Sector
Real Decreto 1221/2010 [114]	1 October 2010	Establishes the procedure of resolution of restrictions by security of supply and amending Royal Decree 2019/1997, of 26th December, which organizes and regulates the electricity production market
Real Decreto 1565/2010 [115]	19 November 2010	Regulates and modifies certain aspects relating to the activity of production of electrical energy in special regime
Real Decreto 1614/2010 [116]	7 December 2010	Regulates and modifies certain aspects relating to the activity of production of electrical energy from technologies solar thermoelectric power and wind power.
Real Decreto -Ley 14/2010 [117]	23 December 2010	Urgent measurements are established for the correction of the tariff deficit of the electrical sector.
Real Decreto 1699/2011 [118]	18 November 2011	Regulates the connection to network of production facilities of electrical energy of small power.
Real Decreto -Ley 1/2012 [119]	27 January 2012	Proceeds to the suspension of the procedures of preallocation of compensation and to the suppression of the economic incentives for new facilities of production of electric power from cogeneration, renewable energy sources and residues

Norm	Date	Ambit
Real Decreto -Ley 2/2013 [120]	1 February 2013	Urgent measures in the electrical system and in the financial sector.
Real Decreto -Ley 9/2013 [121]	12 July 2013	Urgent measurements are adopted to guarantee the financial stability of the electrical system.
Ley 24/2013 [86]	26 December 2013	The electricity sector.

Appendix C.

The main application standards for the construction and development to take into account for a SMES device are found in Table C.1.

Table C.1. Main standards relative to the SMES systems [105]

Norm	Ambit	European Equivalent	International Equivalent	CTN
UNE-EN 286-1:1999	Simple pressure receptacles not submitted to the flame, designed to contain air or nitrogen. Part 1: Pressure receptacles for general uses.	EN 286-1:1998		AEN/CTN 62
UNE-EN 286-1/A1:2003	Simple pressure receptacles not submitted to the flame, designed to contain air or nitrogen. Part 1: Pressure receptacles for general uses.	EN 286-1:1998/AC:2002; EN 286-1:1998/A1:2002		AEN/CTN 62
UNE-EN 286-1:1999/A2:2006	Simple pressure receptacles not submitted to the flame, designed to contain air or nitrogen. Part 1: Pressure receptacles for general uses.	EN 286-1:1998/A2:2005		AEN/CTN 62
UNE-EN 13371:2002	Cryogenic receptacles. Couplings for cryogenic use.	EN 13371:2001		AEN/CTN 62
UNE-EN 13275:2001	Cryogenic receptacles. Pumps for cryogenic use.	EN 13275:2000		AEN/CTN 62
UNE-EN 1797:2002	Cryogenic receptacles. Gas/material compatibility	EN 1797:2001		AEN/CTN 62

Table C.1. (Continued)

Norm	Ambit	European Equivalent	International Equivalent	CTN
UNE-EN 13648-1:2009	Cryogenic receptacles. Safety devices for protection against excessive pressure. Part 1: Safety valves for the cryogenic service	EN 13648-1:2008		AEN/CTN 62
UNE-EN 13648-2:2002	Cryogenic receptacles. Safety devices for protection against excessive pressure. Part 2: Safety Devices with rupture disks for the cryogenic service	EN 13648-2:2002		AEN/CTN 62
UNE-EN 13648-3:2003	Cryogenic receptacles. Safety devices for protection against excessive pressure. Part 3: Determination of the required discharge. Capacity and sizing	EN 13648-3:2002		AEN/CTN 62
UNE-EN 13530-1:2002	Cryogenic receptacles Big transportable receptacles isolated in vacuum. Part 1: Fundamental requirements.	EN 13530-1:2002		AEN/CTN 62
UNE-EN 13530-2:2003	Cryogenic receptacles Big transportable receptacles isolated in vacuum. Part 2: Design, fabrication, inspection and testing.	EN 13530-2:2002		AEN/CTN 62
UNE-EN 13530-2:2003/AC: 2007	Cryogenic receptacles Big transportable receptacles isolated in vacuum. Part 2: Design, fabrication, inspection and testing.	EN 135302:2002/AC:2006		AEN/CTN 62
UNE-EN 13530-2/A1:2004	Cryogenic receptacles Big transportable receptacles isolated in vacuum. Part 2: Design, fabrication, inspection and testing	EN 13530-2:2002/A1:2004		AEN/CTN 62
UNE-EN 13530-3:2002/A1:2005	Cryogenic receptacles Big transportable receptacles isolated in vacuum. Part 3: Operating Requirements.	EN 13530-3:2002/A1:2005		AEN/CTN 62
UNE-EN 13530-3:2002	Cryogenic receptacles Big transportable receptacles isolated in vacuum. Part 3: Operating Requirements	EN 13530-3:2002		AEN/CTN 62
UNE-EN 14398-1:2004	Cryogenic receptacles Big transportable receptacles non isolated in vacuum. Part 1: Fundamental requirements.	EN 14398-1:2003		AEN/CTN 62
UNE-EN 14398-2:2004+A2:2008	Cryogenic receptacles Big transportable receptacles non isolated in vacuum. Part 2: Design, fabrication, inspection and testing.	EN 14398-2:2003+A2:2008		AEN/CTN 62

Norm	Ambit	European Equivalent	International Equivalent	CTN
UNE-EN 14398-3:2004	Cryogenic receptacles Big transportable receptacles non isolated in vacuum. Part 3: Operating Requirements	EN 14398-3:2003		AEN/CTN 62
UNE-EN 14398-3:2004/A1:2005	Cryogenic receptacles Big transportable receptacles non isolated in vacuum. Part 3: Operating Requirements.	EN 14398-3:2003/A1:2005		AEN/CTN 62
UNE-EN 12300:1999	Cryogenic receptacles. Cleaning for cryogenic service.	EN 12300:1998		AEN/CTN 62
UNE-EN 12300:1999/A1:2006	Cryogenic receptacles Cleaning for cryogenic service.	EN 12300:1998/A1:2006		AEN/CTN 62
UNE-EN 12434:2001	Cryogenic receptacles. Cryogenic flexible hoses.	EN 12434:2000; EN 12434:2000/AC:2001		AEN/CTN 62
UNE-EN 1252-1:1998	Cryogenic receptacles. Materials. Part 1: Requirements of tenacity for temperature below - 80 ° c.	EN 1252-1:1998		AEN/CTN 62
UNE-EN 1252-1/AC:1999	Cryogenic receptacles. Materials. Part 1: Requirements of tenacity for temperature below - 80 ° c	EN 1252-1:1998/AC:1998		AEN/CTN 62
UNE-EN 1252-2:2002	Cryogenic receptacles. Materials. Part 2: Requirements of tenacity to temperatures ranging from - 80 ° C and - 20 ° C.	EN 1252-2:2001		AEN/CTN 62
UNE-EN 12213:1999	Cryogenic receptacles. Evaluation methods of the yield of the isolation.	EN 12213:1998		AEN/CTN 62
UNE-EN 13458-1:2002	Cryogenic receptacles. Static vacuum insulated receptacles. Part 1: Fundamental requirements.	EN 13458-1:2002		AEN/CTN 62
UNE-EN 13458-2:2003	Cryogenic receptacles Static vacuum insulated receptacles. Part 2: Design, fabrication, inspection and testing	EN 13458-2:2002		AEN/CTN 62
UNE-EN 13458-2:2003/AC:2007	Cryogenic receptacles Static vacuum insulated receptacles.. Part 2: Design, fabrication, inspection and testing	EN 134582:2002/AC:2006		AEN/CTN 62
UNE-EN 13458-3:2003	Cryogenic receptacles Static vacuum insulated receptacles. Part 3: Operating Requirements	EN 13458-3:2003		AEN/CTN 62

Table C.1. (Continued)

Norm	Ambit	European Equivalent	International Equivalent	CTN
UNE-EN 13458-3:2003/A1:2005	Cryogenic receptacles Static vacuum insulated receptacles. Part 3: Operating Requirements	EN 13458-3:2003/A1:2005		AEN/CTN 62
UNE-EN 14197-1:2004	Cryogenic receptacles Static non vacuum insulated receptacles Part 1: Fundamental requirements.	EN 14197-1:2003		AEN/CTN 62
UNE-EN 14197-2:2004/A1:2006	Cryogenic receptacles Static non-vacuum insulated receptacles. Part 2: Design, fabrication, inspection and testing	EN 14197-2:2003/A1:2006		AEN/CTN 62
UNE-EN 14197-2:2004	Cryogenic receptacles. Static non-vacuum insulated receptacles. Part 2: Design, fabrication, inspection and testing	EN 14197-2:2003		AEN/CTN 62
UNE-EN 14197-2:2004/AC:2007	Cryogenic receptacles. Static non-vacuum insulated receptacles. Part 2: Design, fabrication, inspection and testing	EN 14197-2:2003/AC:2006		AEN/CTN 62
UNE-EN 14197-3/AC:2004	Cryogenic receptacles. Static non-vacuum insulated receptacles. Part 3: Operating Requirements	EN 14197-3:2004/AC:2004		AEN/CTN 62
UNE-EN 14197-3:2004	Cryogenic receptacles Static non-vacuum insulated receptacles. Part 3: Operating Requirements	EN 14197-3:2004		AEN/CTN 62
UNE-EN 14197-3:2004/A1:2005	Cryogenic receptacles. Static non-vacuum insulated receptacles. Part 3: Operating Requirements	EN 141973:2004/A1:2005		AEN/CTN 62
UNE-EN 1251-1:2001	Cryogenic receptacles Portable receptacles vacuum isolated, not more than 1000 litres volume. Part 1: Fundamental requirements.	EN 1251-1:2000		AEN/CTN 62
UNE-EN 1251-2:2001	Cryogenic receptacles Portable receptacles vacuum isolated, not more than 1000 litres volume. Part 2: Design, fabrication, inspection and testing	EN 1251-2:2000		AEN/CTN 62
UNE-EN 1251-2:2001/AC:2007	Cryogenic receptacles Portable receptacles vacuum isolated, not more than 1000 litres volume. Part 2: Design, fabrication, inspection and testing	EN 1251-2:2000/AC:2006		AEN/CTN 62

Norm	Ambit	European Equivalent	International Equivalent	CTN
UNE-EN ISO 21029-2:2016	Cryogenic receptacles Portable receptacles vacuum isolated, not more than 1000 litres volume. Part 2: Operating Requirements	EN ISO 21029-2:2015	ISO 21029-2:2015	AEN/CTN 62
UNE-EN 1626:2009	Cryogenic receptacles. Valves for cryogenic services	EN 1626:2008		AEN/CTN 62
UNE-EN 61788-1:2007	Superconductivity Part 1: Measurement of the critical current. Continuous critical current of superconductors consisted of the type Cu/Nb-Ti (Ratified by AENOR in April 2007)	EN 61788-1:2007	IEC 61788-1:2006	AEN/CTN 206
UNE-EN 61788-10:2007	Superconductivity Part 10: Measurement of the critical temperature. Critical temperature of the superconductors composed by a method of resistance.	EN 61788-10:2006	IEC 61788-10:2006	AEN/CTN 206
UNE-EN 61788-11:2011	Superconductivity Part 11: Measurement of the relation of residual resistance. Relation of residual resistance of compound superconductors of Nb_3Sn. (Ratified by AENOR in November 2011)	EN 61788-11:2011	IEC 61788-11:2011	AEN/CTN 206
UNE-EN 61788-12:2004	Superconductivity Part 12: Measurement of the relation between matrix and superconductor volumes. Relation between volumes of copper and the rest of the threads compound superconductors of Nb_3Sn.	EN 61788-12:2002	IEC 61788-12:2002	AEN/CTN 206
UNE-EN 61788-12:2013	Superconductivity Part 12: Measurement of the relation between matrix and superconductor volumes. Relation between volumes of copper and the rest of the threads compound superconductors of Nb_3Sn. (Ratified by AENOR in November 2013)	EN 61788-12:2013	IEC 61788-12:2013	AEN/CTN 206
UNE-EN 61788-13:2012	Superconductivity Part 13: Measurement of losses in alternating current. Methods of measurement for magnetometer compounds hysteresis losses in superconducting multifilaments (Ratified by AENOR in November 2012)	EN 61788-13:2012	IEC 61788-13:2012	AEN/CTN 206

Table C.1. (Continued)

Norm	Ambit	European Equivalent	International Equivalent	CTN
UNE-EN 61788-14:2010	Superconductivity Part 14: Superconductors of power devices. General requirements for the testing of characterization of the current cables designed to feed the superconductor devices (Ratified by AENOR in November 2010)	EN 61788-14:2010	IEC 61788-14:2010	AEN/CTN 206
UNE-EN 61788-15:2011	Superconductivity. Part 15: Measurement of the electronic characteristics. Impedance of the intrinsic surface of superconductive movies to the microwave frequencies. (Ratified by AENOR in March 2012)	EN 61788-15:2011	IEC 61788-15:2011	AEN/CTN 206
UNE-EN 61788-16:2013	Superconductivity Part 16: Measures of electronic characteristics. Surface resistance dependent on the power of superconductors at microwave frequencies (Ratified by AENOR in May 2013)	EN 61788-16:2013	IEC 61788-16:2013	AEN/CTN 206
UNE-EN 61788-17:2013	Superconductivity Part 17: Measurements of the electronic characteristics. Local critical current density and its distribution in superconductive movies of big surface. (Ratified by AENOR in May 2013)	EN 61788-17:2013	IEC 61788-17:2013	AEN/CTN 206
UNE-EN 61788-18:2013	Superconductivity Part 18: Measurement of the mechanical properties. Tensile Test at ambient temperature superconductors compounds of BI-2223 and BI-2212 with silver covering. (Ratified by AENOR in January 2014)	EN 61788-18:2013	IEC 61788-18:2013	AEN/CTN 206
UNE-EN 61788-19:2014	Superconductivity Part 19: Measurement of the mechanical properties. Tensile test at ambient temperature of superconductors compound of Nb3Sn in reaction (Ratified by AENOR in March 2014)	EN 61788-19:2014	IEC 61788-19:2013	AEN/CTN 206
UNE-EN 61788-2:2007	Superconductivity Part 2: Measurement of the critical current. Continuous critical current of superconductors compound of Nb_3Sn type (Ratified by AENOR in April 2007)	EN 61788-2:2007	IEC 61788-2:2006	AEN/CTN 206

Norm	Ambit	European Equivalent	International Equivalent	CTN
UNE-EN 61788-21:2015	Superconductivity. Part 21: Superconducting wires. Test methods for practical use of superconducting wires. Guidelines and General characteristics (Ratified by AENOR in August 2015)	EN 61788-21:2015	IEC 61788-21:2015	AEN/CTN 206
UNE-EN 61788-3:2006	Superconductivity Part 3: Measurement of the critical current. Continuous critical current of superconductors oxides of Bi-2212 and Bi-2223 with silver covering (Ratified by AENOR in November 2006)	EN 61788-3:2006	IEC 61788-3:2006	AEN/CTN 206
UNE-EN 61788-4:2016	Superconductivity. Part 4: Measurement of the residual resistance ratio. Relation of residual strength of superconductors compound of Nb-Ti y Nb_3Sn. (Ratified by AENOR in May 2016)	EN 61788-4:2016	IEC 61788-4:2016	AEN/CTN 206
UNE-EN 61788-4:2011	Superconductivity. Part 4: Measurement of the residual resistance ratio. Relation of residual strength of superconductors compound of Nb-Ti. (Ratified by AENOR in November 2011)	EN 61788-4:2011	IEC 61788-4:2011	AEN/CTN 206
UNE-EN 61788-5:2013	Superconductivity Part 5: Measurement of the relation between matrix and superconductor volumes. Relation between volumes of copper and of superconductor cables compound of Cu/Nb-Ti. (Ratified by AENOR in October 2013)	EN 61788-5:2013	IEC 61788-5:2013	AEN/CTN 206
UNE-EN 61788-5:2002	Superconductivity Part 5: Measurement of the relation between matrix and superconductor volumes. Relation between volumes of copper and of superconductor cables compound of Cu/Nb-Ti.	EN 61788-5:2001	IEC 61788-5:2000	AEN/CTN 206
UNE-EN 61788-6:2011	Superconductivity Part 6: Measurement of the mechanical properties. Tensile Test at ambient temperature of superconductors compounds of Cu/Nb-Ti. (Ratified by AENOR in November 2011)	EN 61788-6:2011	IEC 61788-6:2011	AEN/CTN 206
UNE-EN 61788-7:2006	Superconductivity Part 7: Measurement of the electronic properties. Surface resistance of superconductors at microwave frequencies. (Ratified by AENOR in April 2007)	EN 61788-7:2006	IEC 61788-7:2006	AEN/CTN 206

Table C.1. (Continued)

Norm	Ambit	European Equivalent	International Equivalent	CTN
UNE-EN 61788-8:2010	Superconductivity Part 8: Measures of losses in alternating current. Measure through detection coils of total losses in alternating current of the superconductor wires of circular section exposed to a magnetic transverse alternate field from the temperature of liquid helium. (Ratified by AENOR in March 2011)	EN 61788-8:2010	IEC 61788-8:2010	AEN/CTN 206
UNE-EN 61788-9:2005	Superconductivity Part 9: Measures for solid superconductors of high temperature. Density of residual flow of oxides superconductors of bulk grain. (Ratified by AENOR in November 2005)	EN 61788-9:2005	IEC 61788-9:2005	AEN/CTN 206
UNE 21302-815:2001	Electrotechnical vocabulary. Chapter 815. Superconductivity.		IEC 60050-815:2000	AEN/CTN 191
UNE 21302-482:2005	Electrotechnical vocabulary. Part 482: Batteries and electric accumulators.		IEC 60050-482:2004	AEN/CTN 191

Appendix D.

D.1. United States of America

In the USA, the normative elements of the electrical system are structured in hierarchical levels, which implies that the energy policy of the United States is fundamentally determined by State and federal public entities. Energy policy may include legislation, international treaties, subsidies and investment incentives, advice for saving energy, taxes and other public policy techniques. The main law in the U.S. electrical system is the Energy Policy Act of 2005 PL 109-58 [97], which regulates the electric system. The rest of the rules and regulations at the federal level depends on this law.

The federal agencies are obliged to comply with the orders of the administration of energy that, apart from the indicated law, include the following federal laws:

- Executive Order 13693—Planning for Federal Sustainability in the Next Decade [122].
- Energy Independence and Security Act of 2007 [123].
- Executive Order 13221—Energy Efficient Standby Power Devices [124].
- Energy Policy Act of 1992 [125].
- National Energy Conservation Policy Act.

D.2. Japan

The Japanese energy policy is based on the Basic Law of Politics of Energy, which came into force in June, 2002, Law number 71, and it is possible to summarise by the trilemma of "3 E": the energy safety (article 2), the environment sustainability (article 3) and the economic efficiency (article 4) [126, 127]. The Basic Law does not establish quantifiable targets, but it authorizes the government to formulate a strategic plan of energy that promotes measurements to guarantee an energy supply that satisfies the needs for the demand.

The First Strategic Plan of Energy dates of 2003 and since then it has been checked on three occasions: 2007, 2010 and 2014.

With the Third Strategic Plan of Energy, economic efficiency and energy security were subordinate to the "E" of the environment. This plan supported the forecasts from an energy mix in which nuclear energy (in quality of clean energy, efficient and economical) was called to play a leading role, and renewable energies would complement it.

This Plan was valid at the beginning of 2011, at the time of the Fukushima nuclear accident. However, after the accident of Fukushima, the government took a radical turn to aim at the total abandonment of the nuclear energy model. This rotation is materialized in the Innovative Strategy for Energy and the Environment of 2012.

The Innovative Strategy sought to reduce the dependence of both nuclear energy and fossil fuels, maximizing the "green energy" and enhancing the energy efficiency and the renewable energies. The new strategy also reviewed the objectives for CO_2 emissions for 2030.

A White Paper on Energy 2013 was published in June 2014, preceded in March 2014 by the Fourth Strategic Plan of Power [98] (enerugi kihon keikaku) with a horizon of 2030 without specifying the future energy mix in Japan.

In regard to the lines of the Fourth Strategic Plan of Energy, the new energy policy of Japan aims to simultaneously reduce the costs of generation and purchase of primary energy, distribution and consumption, paving the way for the return of nuclear energy.

D.3.. Germany

Germany has a hierarchical legislative structure, in which the first level is the federal government followed by the 16 states that compose Germany, called Länder or Bundesländer, as well as subdivisions of these.

At the federal level, the law of power supply (Stromeinspeisungsgesetz) entered into force in 1991 [128]. For the first time the obligation of the big electrical companies to buy electric power generated with renewable conversion processes was regulated, and they have to pay for it at tariffs previously established. This greatly facilitates the access of "green electricity" to the networks [99].

In the year 2000 the Law of Renewable Energies (EEG) entered into force. With the EEG is enshrined the priority of electricity from renewable energy sources and the connection to the network. The EEG is transformed from then on engine for the development of renewable energies, among other reasons, owing to the regulatory framework. Since the year 2000 the EEG has already been subjected to several amendments: EEG 2004, EEG 2009, EEG 2012 and EGG 2014.

It is in this last reform of the Law [100] that it is intended to increase the energy capacity. Renewable energies and converted energy storage is a key aspect for the future. The main objective is to balance the problems of flashing that the renewable energies created in the electrical system.

The German authorities have opted for the storage of water by pumping as a solution to the energy storage. But the research and development of new ESS, as hybrid systems, have increased for the development of the German electrical system.

Part of the amendments that have been mentioned, the reform of the Law on renewable energy, called EEG 2017, entered into force on 1 January 2017. With this reform, the premium is not fixed by the State, but through market auctions, which depend on the type of renewable energy, with an annual amount being fixed for each one. The aim is to increase the share of renewable energies, from the current 33% to 40-45% in 2025 and to 55-60% in 2035.

REFERENCES

[1] Peck P, Parker T. The 'Sustainable Energy Concept' e making sense of norms and co-evolution within a large research facility's energy strategy. *Journal of Cleaner Production* 2016; 123: 137-154.

[2] Shivarama-Krishna K, Sathish Kumar K. A review on hybrid renewable energy systems. *Renewable and Sustainable Energy Reviews* 2015; 52: 907–916.

[3] Aneke M, Wang M. Energy storage technologies and real life applications – A state of the art review. *Applied Energy* 2016; 179: 350–377.

[4] Global EV Outlook 2015, *IEA*, Paris. https://www.iea.org/ [Accessed on 05/07/2017].

[5] *Global Energy Storage Database, Sandia National Laboratories.* http://www.energystorageexchange.org/ [Accedido el 05/07/2017].

[6] REE. El Sistema Eléctrico Español, Informe 2015 [The Spanish Electric System, 2015 Report]. *Red Eléctrica de España (REE).* http://www.ree.es/es/estadisticas-del-sistema-electrico-espanol/informe-anual/informe-del-sistema-electrico-espanol-2015 [Accessed on 05/07/2017].

[7] Palizban O, Kauhaniemi K. Energy storage systems in modern grids—Matrix of technologies and applications. *Journal of Energy Storage* 2016; 6: 248–259.

[8] Kousksou T, Bruel P, Jamil A, El Rhafiki T, Zeraouli Y. Energy storage: Applications and challenges. *Solar Energy Materials & Solar Cells* 2014; 120: 59–80.

[9] Li Y, Li Y, Ji P, Yang J. Development of energy storage industry in China: A technical and economic point of review. *Renewable and Sustainable Energy Reviews* 2015; 49: 805–812.

[10] Zhu J et al. Design, dynamic simulation and construction of a hybrid HTS SMES (high-temperature superconducting magnetic energy storage systems) for Chinese power grid. *Energy* 2013; 51: 184-192.

[11] Saboori H, Hemmati R, Jirdehi MA. Reliability improvement in radial electrical distribution network by optimal planning of energy storage systems. *Energy* 2015; 93: 2299-2312.

[12] Ould-Amrouche S, Rekioua D, Rekioua T, Bacha S. Overview of energy storage in renewable energy systems. *International Journal of Hydrogen Energy* 2016; 45: 20914–20927.

[13] Jin JX, Chen XY. Study on the SMES Application Solutions for Smart Grid. *Physics Procedia* 2012; 36: 902 – 907.

[14] Aly MM, Abdel-Akher M, Said SM, Senjyu T. A developed control strategy for mitigating wind power generation transients using superconducting magnetic energy storage with reactive power support. *Electrical Power and Energy Systems* 2016; 83: 485–494.

[15] Farhadi-Kangarlu M, Alizadeh-Pahlavani MR. Cascaded multilevel converter based superconducting magnetic energy storage system for frequency control. *Energy* 2014; 70: 504-513.

[16] Zhu J et al. Experimental demonstration and application planning of high temperature superconducting energy storage system for renewable power grids. *Applied Energy* 2015; 137: 692–698.

[17] Hasan NS, Hasan MY, Majid MS, Rahman HA. Review of storage schemes for wind energy systems. *Renewable and Sustainable Energy Reviews* 2013; 21: 237–247.

[18] Hirano N, Watanabe T, Nagaya S. Development of cooling technologies for SMES. *Cryogenics* 2016; 80: 210–214.

[19] Dargahi V, Sadigh AK, Pahlavani MRA, Shoulaie A. DC (direct current) voltage source reduction in stacked multicell converter based energy systems. *Energy* 2012; 46: 649-663.

[20] Tan X, Li Q, Wang H. Advances and trends of energy storage technology in Microgrid. *Electrical Power and Energy Systems* 2013; 44: 179–191.

[21] Mariam L, Basu M, Conlon, MF. Microgrid: Architecture, policy and future trends. *Renewable and Sustainable Energy Reviews* 2016; 64: 477–489.

[22] Colmenar-Santos A, Linares-Mena AR, Velazquez JF, Borge-Diez D. Energy-efficient three-phase bidirectional converter for grid-connected storage applications. *Energy Conversion and Management* 2016; 127: 599–611.

[23] Castillo A, Gayme DF. Grid-scale energy storage applications in renewable energy integration: A survey. *Energy Conversion and Management* 2014; 87: 885–894.

[24] Luo X, Wang J, Dooner M, Clarke J. Overview of current development in electrical energy storage technologies and the application potential in power system operation. *Applied Energy* 2015; 137: 511–536.

[25] Zheng M, Meinrenken CJ, Lackner KS. Smart households: Dispatch strategies and economic analysis of distributed energy storage for residential peak shaving. *Applied Energy* 2015; 147: 246–257.

[26] Chatzivasileiadi A, Ampatzi E, Knight I. Characteristics of electrical energy storage technologies and their applications in buildings. *Renewable and Sustainable Energy Reviews* 2013; 25: 814–830.

[27] Ferreira HL, Garde R, Fulli G, Kling W, Lopes JP. Characterisation of electrical energy storage technologies. *Energy* 2013; 53: 288-298.

[28] Evans A, Strezov V, Evans TJ. Assessment of utility energy storage options for increased renewable energy penetration. *Renewable and Sustainable Energy Reviews* 2012; 16: 4141– 4147.

[29] Rodrigues EMG et al. Energy storage systems supporting increased penetration of renewables in islanded systems. *Energy* 2014; 75: 265-280.

[30] Ren L et al. Techno-economic evaluation of hybrid energy storage technologies for a solar–wind generation system. *Physica* C 2013; 484: 272–275.

[31] Gaudard L, Romerio F. Reprint of "The future of hydropower in Europe: Interconnecting climate, markets and policies". *Environmental Science & Policy* 2014; 37: 172–181.

[32] Gallo AB, Simoes-Moreira JR, Costa HKM, Santos MM, Moutinho-dos-Santos E. Energy storage in the energy transition context: A technology review. *Renewable and Sustainable Energy Reviews* 2016; 65: 800–822.

[33] Lund PD, Lindgren J, Mikkola J, Salpakari J. Review of energy system flexibility measures to enable high levels of variable renewable electricity. *Renewable and Sustainable Energy Reviews* 2015; 45; 785–807.

[34] Hemmati R, Saboori H. Emergence of hybrid energy storage systems in renewable energy and transport applications – A review. *Renewable and Sustainable Energy Reviews* 2016; 65: 11–23.

[35] Solomon AA, Faiman D, Meron G. Appropriate storage for high-penetration grid-connected photovoltaic plants. *Energy Policy* 2012; 40: 335–344.

[36] Theo WL et al. An MILP model for cost-optimal planning of an on-grid hybrid power system for an eco-industrial park. *Energy* 2016; 116: 1423–1441.

[37] Planas E, Andreu J, Gárate JI, Martinez-de-Alegría I, Ibarra E. AC and DC technology in microgrids: A review. *Renewable and Sustainable Energy Reviews* 2015; 43: 726–749.

[38] Mahlia TMI, Saktisahdan TJ, Jannifar A, Hasan MH, Matseelar HSC. A review of available methods and development on energy storage; Technology update. *Renewable and Sustainable Energy Reviews* 2014; 33: 532–545.

[39] Yang J, Liu W, Liu P. Application of SMES unit in Black Start. *Physics Procedia* 2014; 58: 277 – 281.

[40] Selvaraju RK, Somaskandan G. Impact of energy storage units on load frequency control of deregulated power systems. *Energy* 2016; 97: 214-228.

[41] Koohi-Kamali, S. et al. Emergence of energy storage technologies as the solution for reliable operation of smart power systems: A review. *Renewable and Sustainable Energy Reviews* 2013; 25: 135–165.

[42] Kyriakopoulos GL, Arabatzis G. Electrical energy storage systems in electricity generation: Energy policies, innovative technologies, and regulatory regimes. *Renewable and Sustainable Energy Reviews* 2016; 56:1044–1067.

[43] Tuya-Merino M. Simulación eléctrica de líneas ferroviarias electrificadas para el diseño de un sistema de almacenamiento de energía para la recarga de vehículos eléctricos [Electrical simulation of electrified railway lines for the design of an energy storage system for charging electric vehicles]. Escuela Ingenierías Industriales, Depto. *Ingeniería Energética y Fluidomecánica.* http://uvadoc.uva.es/handle/10324/13760 [Accessed on 05/07/2017].

[44] Ministerio de Industria, Energía y Turismo. La Energía en España 2014 [Energy in Spain 2014], http://www.minetad.gob.es/energia/balances/Balances/LibrosEnergia/La_Energ%C3%ADa_2014.pdf [Accessed on 05/07/2017].

[45] *Official website of the European Union.* https://europa.eu/european-union/law/legal-acts_en [Accessed on 05/07/2017].

[46] Constitución española Título III. De las Cortes Generales. Capítulo segundo. *De la elaboración de las leyes.* [Spanish Constitution Title III. Of the General Cortes. Chapter two. Of the elaboration of the laws], http://www.congreso.es/consti/constitucion/indice/titulos/articulos.jsp?ini=81&fin=92&tipo=2 [Accessed on 05/07/2017].

[47] Constitución española Título III. *De las Cortes Generales.* Sinopsis artículo 81 [Spanish Constitution Title III. Of the General Cortes. Synopsis Article 81], http://www.congreso.es/consti/constitucion/indice/sinopsis/sinopsis.jsp?art=81&tipo=2 [Accessed on 05/07/2017].

[48] Yan X, Zhang X, Chen H, Xu Y, Tan C. Techno-economic and social analysis of energy storage for commercial buildings. *Energy Conversion and Management* 2014; 78: 125–136.

[49] Bradbury K, Pratson L, Patiño-Echeverri D. Economic viability of energy storage systems based on price arbitrage potential in real-time U.S. electricity markets. *Applied Energy* 2014; 114: 512–519.

[50] Spataru C, Kok YC, Barrett M, Sweetnam T. Techno-Economic Assessment for Optimal Energy Storage Mix. *Energy Procedia* 2015; 83: 515 – 524.

[51] Landry M, Gagnon Y. Energy Storage: Technology Applications and Policy Options. *Energy Procedia* 2015; 79: 315 – 320.

[52] Sundararagavan S, Baker E. Evaluating energy storage technologies for wind power integration. *Solar Energy* 2012; 86: 2707–2717.

[53] Zheng M, Meinrenken CJ, Lackner KS. Agent-based model for electricity consumption and storage to evaluate economic viability of tariff arbitrage for residential sector demand response. *Applied Energy* 2014; 126: 297–306.

[54] Leou R-C. An economic analysis model for the energy storage system applied to a distribution substation. *Electrical Power and Energy Systems* 2012; 34: 132–137.

[55] Han P, Wu Y, Liu H, Li L, Yang H. Structural design and analysis of a 150 kJ HTS SMES cryogenic System. *Physics Procedia* 2015; 67: 360 – 366.

[56] Hossain J, Mahmud A. *Large Scale Renewable Power Generation: Advances in Technologies for Generation, Transmission and Storage*. Springer 2014. http://dx.doi.org/10.1007/978-981-4585-30-9 [Accessed on 05/07/2017].

[57] Vasquez S, Lukic SM, Galvan E, Franquelo LG, Carrasco JM. Energy Storage Systems for Transport and Grid Applications. *IEEE Transactions on Industrial Electronics* 2010; 12: 3881–3895.

[58] Li J, Gee AM, Zhang M, Yuan W. Analysis of battery lifetime extension in a SMES-battery hybrid energy storage system using a novel battery lifetime model. *Energy* 2015; 86: 175-185.

[59] Nagaya S et al. Development of MJ-class HTS SMES for bridging instantaneous voltage dips. *IEEE Transactions on Applied Superconductivity* 2004; 2: 770–773.

[60] Rogers JD, Schermer RI, Miller BL, Hauer JF. 30-MJ superconducting magnetic energy storage system for electric utility transmission stabilization. *Proceedings of the IEEE* 1983; 9: 1099–1107.

[61] Kreutz R et al. Design of a 150 kJ high-Tc SMES (HSMES) for a 20 kVA uninterruptible power supply system. *IEEE Transactions on Applied Superconductivity* 2003; 2: 1860–1862.

[62] Zakeri B, Syri S. Electrical energy storage systems: A comparative life cycle cost analysis. *Renewable and Sustainable Energy Reviews* 2015; 42: 569–596.

[63] Akinyele DO, Rayudu RK. Review of energy storage technologies for sustainable power networks. *Sustainable Energy Technologies and Assessments* 2014; 8: 74–91.

[64] Mahto, T, Mukherjee V. Energy storage systems for mitigating the variability of isolated hybrid power system. *Renewable and Sustainable Energy Reviews* 2015; 51: 1564–1577.

[65] Spataru C, Kok YC, Barrett M. Physical energy storage employed worldwide. *Energy Procedia* 2014; 62: 452 – 461.

[66] *Energy Storage Trends and Opportunities in Emerging Markets.* ESMAP, IFC. http://www.ifc.org/wps/wcm/connect/ed6f9f7f-f197-4915-8ab6-56b92d50865d/7151-IFC-EnergyStorage-report.pdf?MOD=AJPERES [Accessed on 05/07/2017].

[67] *Resumen del Presupuesto General Consolidado 2016* [Summary of the 2016 Consolidated General Budget], Zaragoza. http://www.zaragoza.es/ciudad/encasa/hacienda/presupuestos/presupuestos.htm [Accessed on 05/07/2017].

[68] Informe de responsabilidad corporativa, *Resumen* 2015 [Corporate responsibility report, Summary 2015] (2016), REE. http://www.ree.es/es/publicaciones/informe-anual-2015 [Accessed on 05/07/2017].

[69] CNMC. *Acuerdo por el que se remite a la Secretaría de Estado de Energía la propuesta motivada de la cuantía a percibir por cada empresa distribuidora sobre el incentivo o penalización par la reducción de pérdidas en la red de distribución de energía eléctrica para el año 2016* [Agreement by which the motivated proposal of the amount to be received by each distribution company on the incentive or penalty for the reduction of losses in the electric power distribution network for the year 2016 is sent to the Secretary of State for Energy], Comisión Nacional de los Mercados y Competencia (CNMC). https://www.cnmc.es/sites/default/files/1553612.pdf [Accessed on 05/07/2017].

[70] Ministerio de Energía, Turismo y Agenda Digital, http://www.minetad.gob.es/es-ES/Paginas/index.aspx [Accessed on 05/07/2017].

[71] Estadísticas del Sistema Eléctrico. *Series estadísticas del sistema eléctrico español* (febrero 2017) [*Statistical series of the Spanish electricity system*], http://www.ree.es/es/estadisticas-del-sistema-electrico-espanol/indicadores-nacionales/series-estadisticas [Accessed on 05/07/2017].

[72] European Commission. *Strategy 2020.* http://ec.europa.eu/europe2020/index_en.htm [Accessed on 05/07/2017].

[73] Resolución del Parlamento Europeo, *de 5 de febrero de 2014, sobre un marco para las políticas de clima y energía en 2030* [European Parliament resolution of 5 February 2014 on a framework for climate and energy policies in 2030]. http://www.europarl.europa.eu/sides/getDoc.do?pubRef=-//EP// TEXT+TA+P7-TA-2014-0094+0+DOC+XML+V0//ES [Accessed on 05/07/2017].

[74] Resolución del Parlamento Europeo, *de 14 de marzo de 2013, sobre la Hoja de Ruta de la Energía para 2050, un futuro con energía* [European Parliament resolution of 14 March 2013 on the Energy Roadmap for 2050, a future with energy.] http://www.europarl.europa.eu/sides/getDoc.do?pubRef=-//EP//TEXT+TA+P7-TA-2013-0088+0+DOC+XML+V0//ES [Accessed on 05/07/2017].

[75] Resolución del Parlamento Europeo, *de 21 de mayo de 2013, sobre los desafíos y oportunidades actuales para las energías renovables*

en el mercado interior europeo de la energía [*European Parliament resolution of 21 May 2013 on current challenges and opportunities for renewable energies in the European internal energy market*]. http://www.europarl.europa.eu/sides/getDoc.do?pubRef=-//EP//TEXT+TA+P7-TA-2013-0201+0+DOC+XML+V0//ES [Accessed on 05/07/2017].

[76] Resolución del Parlamento Europeo, *de 25 de noviembre de 2010, sobre una nueva estrategia energética para Europa 2011-2020* [European Parliament resolution of 25 November 2010 on a new energy strategy for Europe 2011-2020] (DO C 99 E de 3.4.2012, p. 64). http://www.europarl.europa.eu/sides/getDoc.do?pubRef=-//EP//TEXT+TA+P7-TA-2010-0441+0+DOC+XML+V0//ES [Accessed on 05/07/2017].

[77] Resolución del Parlamento Europeo, *de 5 de julio de 2011, sobre las prioridades de la infraestructura energética a partir de 2020* [European Parliament resolution of 5 July 2011 on the priorities of the energy infrastructure after 2020] (DOC 33 E de 5.2.2012, p. 46). http://eur-lex.europa.eu/legal-content/ES/TXT/?uri=CELEX% 3A52011IP0318 [Accessed on 05/07/2017].

[78] Resolución del Parlamento Europeo, de 12 de septiembre de 2013, sobre la microgeneración - generación de electricidad y de calor a pequeña escala [European Parliament resolution of 12 September 2013 on micro-generation - generation of electricity and heat on a small scale] http://www.europarl.europa.eu/sides/getDoc.do?pubRef=-//EP//TEXT+TA+P7-TA-2013-0374+0+DOC+XML+V0//ES [Accessed on 05/07/2017].

[79] Grid+Storage consortium. http://www.gridplusstorage.eu/ [Accessed on 05/07/2017].

[80] Title Xxi, Energy, Article 194, Consolidated Versions of the Treaty on European Union and the Treaty on the Functioning of the European Union. Charter of Fundamental Rights of the European Union, 2010. http://eur-lex.europa.eu/legal-content/EN/TXT/?uri=celex%3A12012E%2FTXT%20 [Accessed on 05/07/2017].

[81] Directive 2009/28/EC of the European Parliament and of the Council, 23 April 2009. http://eur-lex.europa.eu/legal-content/EN/TXT/PDF/?uri=CELEX:32009L0028 [Accessed on 05/07/2017].
[82] Directive 2009/72/CE of the European Parliament and of the Council, 13 July 2009. https://www.boe.es/doue/2009/211/L00055-00093.pdf [Accessed on 05/07/2017].
[83] Directive 2012/27/EU of the European Parliament and of the Council, 25 October 2012. http://eur-lex.europa.eu/legal-content/EN/TXT/PDF/?uri=CELEX:32012L0027%20 [Accessed on 05/07/2017].
[84] Regulation (EU) No 347/2013 of the European Parliament and of the Council, 17 April 2013. http://eur-lex.europa.eu/legal-content/AUTO/?uri=CELEX: 12012E172 [Accessed on 05/07/2017].
[85] Ley 54/1997, de 27 de noviembre, del Sector Eléctrico [Law 54/1997, of November 27, of the Electricity Sector], https://www.boe.es/diario_boe/txt.php?id=BOE-A-1997-25340 [Accessed on 05/07/2017].
[86] Ley 24/2013, *de 26 de diciembre, del Sector Eléctrico* [Law 24/2013, of December 26, of the Electricity Sector]. https://www.boe.es/diario_boe/txt.php?id=BOE-A-2013-13645 [Accessed on 05/07/2017].
[87] Resolución de 30 de julio de 1998, *de la Secretaría de Estado de Energía y Recursos Minerales, por la que se aprueba un conjunto de procedimientos de carácter técnico e instrumental necesarios para realizar la adecuada gestión técnica del sistema eléctrico* [Resolution of July 30, 1998, from the Secretary of State for Energy and Mineral Resources, approving a set of procedures of a technical and instrumental nature necessary to carry out the appropriate technical management of the electrical system]. https://www.boe.es/boe/dias/1998/08/18/pdfs/A28158-28183.pdf [Accessed on 05/07/2017].
[88] Resolución de 17 de marzo de 2004, *de la Secretaría de Estado de Energía, Desarrollo Industrial y Pequeña y Mediana Empresa, por*

la que se modifican un conjunto de procedimientos de carácter técnico e instrumental necesarios para realizar la adecuada gestión técnica del Sistema Eléctrico [Resolution of March 17, 2004, of the Secretary of State for Energy, Industrial Development and Small and Medium Enterprises, whereby a set of procedures of a technical and instrumental nature are modified to carry out the adequate technical management of the Electric System%5d.%20http:/www.ree.es/sites/default/files/01_ACTIVIDADES/Documentos/ProcedimientosOperacion/PO_resol_17mar2004_correc_c.pdf [Accessed on 05/07/2017].

[89] Resolución de 18 de diciembre de 2015, *de la Secretaría de Estado de Energía, por la que se establecen los criterios para participar en los servicios de ajuste del sistema y se aprueban determinados procedimientos de pruebas y procedimientos de operación para su adaptación al Real Decreto 413/2014, de 6 de junio, por el que se regula la actividad de producción de energía eléctrica a partir de fuentes de energía renovables, cogeneración y residuos* [Resolution of December 18, 2015, of the Secretary of State for Energy, which establishes the criteria for participating in the adjustment services of the system and approves certain testing procedures and operating procedures for its adaptation to Royal Decree 413 / 2014, of June 6, which regulates the activity of producing electrical energy from renewable energy sources, cogeneration and waste]. https://www.boe.es/boe/dias/2015/12/19/pdfs/BOE-A-2015-13875.pdf [Accessed on 05/07/2017].

[90] Resolución de 27 de octubre de 2010, *de la Secretaría de Estado de Energía, por la que se aprueban los procedimientos de operación del sistema P.O. 3.10, P.O. 14.5, P.O. 3.1, P.O. 3.2, P.O. 9 y P.O. 14.4 para su adaptación a la nueva normativa eléctrica* [Resolution of October 27, 2010, of the Secretary of State for Energy, approving the operating procedures of the P.O. system. 3.10, P.O. 14.5, P.O. 3.1, P.O. 3.2, P.O. 9 and P.O. 14.4 for its adaptation to the new electrical regulations]. https://www.boe.es/boe/dias/2010/10/28/pdfs/BOE-A-2010-16441.pdf [Accessed on 05/07/2017].

[91] Resolución de 10 de marzo de 2000, *de la Secretaría de Estado de Industria y Energía, por la que se aprueba el procedimiento de operación del sistema (P.O. - 7.4) «Servicio complementario de control de tensión de la red de transporte»* [Resolution of March 10, 2000, of the Secretary of State for Industry and Energy, approving the operating procedure of the system (P.O. - 7.4) "Supplementary service of tension control of the transport network"] https://www.boe.es/boe/dias/2000/03/18/pdfs/A11330-11346.pdf [Accessed on 05/07/2017].

[92] Resolución de 7 de abril de 2006, *de la Secretaría General de Energía, por la que se aprueban los procedimientos de operación 8.1 Definición de las redes operadas y observadas por el Operador del Sistema y 8.2 Operación del sistema de producción y transporte* [Resolution of April 7, 2006, of the General Secretariat of Energy, by which the operating procedures are approved 8.1 Definition of the networks operated and observed by the System Operator and 8.2 Operation of the production and transport system], https://www.boe.es/boe/dias/2006/04/21/pdfs/A15341-15345.pdf [Accessed on 05/07/2017].

[93] Resolución de 28 de abril de 2006, *de la Secretaría General de Energía, por la que se aprueba un conjunto de procedimientos de carácter técnico e instrumental necesarios para realizar la adecuada gestión técnica de los sistemas eléctricos insulares y extrapeninsulares* [Resolution of April 28, 2006, of the General Secretariat of Energy, which approves a set of procedures of a technical and instrumental nature necessary to carry out the adequate technical management of the island and extra-peninsular electrical systems] https://www.boe.es/boe/dias/2006/05/31/pdfs/A20573-20574.pdf [Accessed on 05/07/2017].

[94] Resolución de 22 de marzo de 2005, *de la Secretaría General de la Energía, por la que se aprueba el Procedimiento de Operación 13.1. Criterios de Desarrollo de la Red de Transporte, de carácter técnico e instrumental necesario para realizar la adecuada gestión técnica del Sistema Eléctrico* [Resolution of March 22, 2005, of the General

Secretariat of Energy, approving the Operating Procedure 13.1. Criteria for the Development of the Transport Network, of a technical and instrumental nature necessary to carry out the adequate technical management of the Electric System] http://www.ree.es/sites/default/files/01_ACTIVIDADES/Documentos/ProcedimientosOperacion/PO_resol_22Mar2005.pdf [Accessed on 05/07/2017].

[95] Resolución de 11 de febrero de 2005, *de la Secretaría General de la Energía, por la que se aprueba un conjunto de procedimientos de carácter técnico e instrumental necesarios para realizar la adecuada gestión técnica del Sistema Eléctric*o [Resolution of February 11, 2005, of the General Secretariat of Energy, which approves a set of procedures of a technical and instrumental nature necessary to perform the proper technical management of the Electric System] https://www.boe.es/boe/dias/2005/03/01/pdfs/A07405-07430.pdf [Accessed on 05/07/2017].

[96] *Resolución de 5 de agosto de 2016, de la Secretaría de Estado de Energía, por la que se modifica el Procedimiento de Operación 15.2* Servicio de gestión de la demanda de interrumpibilidad, aprobado por Resolución de 1 de agosto de 2014 [Resolution of August 5, 2016, of the Secretary of State for Energy, by which the Operating Procedure is modified 15.2 Interruptability demand management service, approved by Resolution of August 1, 2014] https://www.boe.es/boe/dias/2016/08/12/pdfs/BOE-A-2016-7800.pdf [Accessed on 05/07/2017].

[97] *Energy Policy Act of 2005 PL 109-58.* https://www.gpo.gov/fdsys/pkg/PLAW-109publ58/pdf/PLAW-109publ58.pdf [Accessed on 05/07/2017].

[98] *Strategic Energy Plan,* April, 2014. http://www.enecho.meti.go.jp/en/category/others/basic_plan/pdf/4th_strategic_energy_plan.pdf [Accessed on 05/07/2017].

[99] *Frankfurter Societäts-Medien GmbH en cooperación con el Ministerio de Relaciones Exteriores de Alemania* [Frankfurter Societäts-Medien GmbH in cooperation with the Ministry of Foreign

Affairs of Germany] https://www.deutschland.de [Accessed on 05/07/2017].

[100] Renewable Energy Sources Act: *Plannable. Affordable. Efficient.* http://www.bmwi.de/English/Redaktion/Pdf/renewable-energy-sources-act-eeg-2014,property=pdf,bereich=bmwi2012,sprache=en,rwb=true.pdf [Accessed on 05/07/2017].

[101] *Conferencia de París sobre el Clima (COP21)* [Paris Conference on Climate (COP21)], http://www.cop21paris.org/ [Accessed on 05/07/2017].

[102] *International Organization for Standardization, ISO.* http://www.iso.org/ [Accessed on 05/07/2017].

[103] *European Committee for Standardization, CEN.* http://www.cen.eu/ [Accessed on 05/07/2017].

[104] *European Committee for Electrotechnical Standardization, CENELEC.* https://www.cenelec.eu/ [Accessed on 05/07/2017].

[105] *Asociación Española de Normalización y Certificación, AENOR* [Spanish Association for Standardization and Certification, AENOR] http://www.aenor.es/aenor/inicio/home/home.asp [Accessed on 05/07/2017].

[106] *Comisión Panamericana de Normas Técnicas, COPANT* [Pan American Standards Commission, COPANT] http://www.copant.org/ [Accessed on 05/07/2017].

[107] *Handbook for Energy Storage for Transmission or Distribution Applications.* Report No. 1007189. Technical Update December 2002. www.epri.com [Accessed on 05/07/2017].

[108] Ter-Gazarian, A., 1994. *Energy Storage for Power Systems.* 1st edition. Peter Peregrinus Ltd.

[109] Real Decreto 2019/1997, *de 26 de diciembre, por el que se organiza y regula el mercado de producción de energía eléctrica* [Royal Decree 2019/1997, of December 26, by which the electric power production market is organized and regulated]. https://www.boe.es/boe/dias/1997/12/27/pdfs/A38047-38057.pdf [Accessed on 05/07/2017].

[110] Real Decreto 1955/2000, *de 1 de diciembre, por el que se regulan las actividades de transporte, distribución, comercialización, suministro y procedimientos de autorización de instalaciones de energía eléctrica* [Royal Decree 1955/2000, of December 1, which regulates the activities of transportation, distribution, marketing, supply and authorization procedures for electric power installations] https://www.boe.es/boe/dias/2000/12/27/pdfs/A45988-46040.pdf [Accessed on 05/07/2017].

[111] Real Decreto-ley 6/2009, *de 30 de abril, por el que se adoptan determinadas medidas en el sector energético y se aprueba el bono social* [Royal Decree-Law 6/2009, of April 30, by which certain measures are adopted in the energy sector and the social bond is approved] https://www.boe.es/boe/dias/2009/05/07/pdfs/BOE-A-2009-7581.pdf [Accessed on 05/07/2017].

[112] Real Decreto 134/2010, *de 12 de febrero, por el que se establece el procedimiento de resolución de restricciones por garantía de suministro y se modifica el Real Decreto 2019/1997, de 26 de diciembre, por el que se organiza y regula el mercado de producción de energía eléctrica* [Royal Decree 134/2010, of February 12, which establishes the procedure of resolution of restrictions for guarantee of supply and modifies Royal Decree 2019/1997, of December 26, by which the market is organized and regulated of electricity production] https://www.boe.es/boe/dias/2010/02/27/pdfs/BOE-A-2010-3158.pdf [Accessed on 05/07/2017].

[113] Real Decreto-ley 6/2010, *de 9 de abril, de medidas para el impulso de la recuperación económica y el empleo* [Royal Decree-Law 6/2010, of 9 April, on measures to boost economic recovery and employment] https://www.boe.es/boe/dias/2010/04/13/pdfs/BOE-A-2010-5879.pdf [Accessed on 05/07/2017].

[114] Real Decreto 1221/2010, *de 1 de octubre, por el que se modifica el Real Decreto 134/2010, de 12 de febrero, por el que se establece el procedimiento de resolución de restricciones por garantía de suministro y se modifica el Real Decreto 2019/1997, de 26 de diciembre, por el que se organiza y regula el mercado de producción*

de energía eléctrica [Royal Decree 1221/2010, of October 1, which modifies Royal Decree 134/2010, of February 12, which establishes the procedure of resolution of restrictions for guarantee of supply and Royal Decree 2019 is modified / 1997, of December 26, by which the electric power production market is organized and regulated] https://www.boe.es/boe/dias/2010/10/02/pdfs/BOE-A-2010-15121.pdf [Accessed on 05/07/2017].

[115] Real Decreto 1565/2010, *de 19 de noviembre, por el que se regulan y modifican determinados aspectos relativos a la actividad de producción de energía eléctrica en régimen especial* [Royal Decree 1565/2010, of November 19, which regulates and modifies certain aspects related to the activity of production of electric power in special regime] https://www.boe.es/boe/dias/2010/11/23/pdfs/BOE-A-2010-17976.pdf [Accessed on 05/07/2017].

[116] Real Decreto 1614/2010, *de 7 de diciembre, por el que se regulan y modifican determinados aspectos relativos a la actividad de producción de energía eléctrica a partir de tecnologías solar termoeléctrica y eólic*a [Royal Decree 1614/2010, of December 7, which regulates and modifies certain aspects related to the activity of production of electric power from solar thermal and wind technologies] https://www.boe.es/boe/dias/2010/12/ 08/pdfs/BOE-A-2010-18915.pdf [Accessed on 05/07/2017].

[117] Real Decreto-ley 14/2010, *de 23 de diciembre, por el que se establecen medidas urgentes para la corrección del déficit tarifario del sector eléctrico* [Royal Decree-Law 14/2010, of December 23, which establishes urgent measures to correct the tariff deficit of the electricity sector] https://www.boe.es/boe/dias/2010/12/24/pdfs/BOE-A-2010-19757.pdf [Accessed on 05/07/2017].

[118] Real Decreto 1699/2011, *de 18 de noviembre, por el que se regula la conexión a red de instalaciones de producción de energía eléctrica de pequeña potencia* [Royal Decree 1699/2011, of November 18, which regulates the connection to the network of small power electric power production facilities], https://www.boe.es/

boe/dias/2011/12/08/pdfs/BOE-A-2011-19242.pdf [Accessed on 05/07/2017].

[119] Real Decreto-ley 1/2012, *de 27 de enero, por el que se procede a la suspensión de los procedimientos de preasignación de retribución y a la supresión de los incentivos económicos para nuevas instalaciones de producción de energía eléctrica a partir de cogeneración, fuentes de energía renovables y residuos* [Royal Decree-Law 1/2012, of January 27, which proceeds to the suspension of the procedures for pre-allocation of remuneration and the elimination of economic incentives for new installations for the production of electricity from cogeneration, sources of renewable energy and waste]. https:// www.boe.es/boe/dias/2012/01/28/ pdfs/BOE-A-2012-1310.pdf [Accessed on 05/07/2017].

[120] Real Decreto-ley 2/2013, *de 1 de febrero, de medidas urgentes en el sistema eléctrico y en el sector financier* [Royal Decree-Law 2/2013, of 1 February, on urgent measures in the electricity system and in the financial sector] https://www.boe.es/boe/dias/2013/02/02/pdfs/BOE-A-2013-1117.pdf [Accessed on 05/07/2017].

[121] Real Decreto-ley 9/2013, *de 12 de julio, por el que se adoptan medidas urgentes para garantizar la estabilidad financiera del sistema eléctrico* [Royal Decree-Law 9/2013, of July 12, by which urgent measures are adopted to guarantee the financial stability of the electrical system] https://www.boe.es/boe/dias/2013/07/13/pdfs/BOE-A-2013-7705.pdf [Accessed on 05/07/2017].

[122] Executive Order 13693—*Planning for Federal Sustainability in the Next Decade.* https://www.gpo.gov/fdsys/pkg/FR-2015-03-25/pdf/2015-07016.pdf [Accessed on 05/07/2017].

[123] *Energy Independence and Security Act of 2007 PL 110-140.* https://www.gpo.gov/fdsys/pkg/BILLS-110hr6enr/pdf/BILLS-110hr6enr.pdf [Accessed on 05/07/2017].

[124] *Executive Order 13221—Energy Efficient Standby Power Devices.* https://energy.gov/sites/prod/files/2013/10/f3/eo13221.pdf [Accessed on 05/07/2017].

[125] *Energy Policy Act of 1992* PL 102-486. http://www.afdc.energy.gov/pdfs/2527.pdf [Accessed on 05/07/2017].

[126] Casado MF. El futuro energético de Japón: entre el regreso a la senda nuclear y el giro hacia las renovables [Japan's energy future: between the return to the nuclear path and the turn towards renewable energies]. *UNISCI Journal* 2016; 41.

[127] *Ministry of Economy, Trade and Industry.* http://www.meti.go.jp/english/index.html [Accessed on 05/07/2017].

[128] *Ley de alimentación de energía eléctrica* (Stromein-speisungsgesetz) [Electric power supply law], http://dip21.bundestag.de/dip21/btd/11/078/1107816.pdf [Accessed on 05/07/2017].

In: Renewable Electric Power Distribution … ISBN: 978-1-53614-202-0
Editors: A. Colmenar-Santos et al.

Chapter 2

SUPERCONDUCTING FAULT CURRENT LIMITER IN DISTRIBUTION SUBSTATIONS

J. M. Pecharromán-Lázaro[1,*], Carlos de Palacio[2] and Manuel Castro-Gil[1]

[1]Departamento de Ingeniería Eléctrica, Electrónica, Control, Telemática y Química Aplicada a la Ingeniería, Universidad Nacional de Educación a Distancia (UNED), Madrid, Spain

[2]ABB, Madrid, Spain

ABSTRACT

All substations that make up an electric energy distribution system are vulnerable to unwanted voltage or current surges, with negative implications for the affected facilities. The consequence are economic losses for the companies which operate them as well as interruption of energy delivery. This chapter presents the implementation of an advanced fault current limiting device based on second generation superconducting.

[*] Corresponding Author Email: jm.pecharroman.lazaro@gmail.com.

The purpose of the system is to avoid the negative effects of network incidents. The results show a reduction of peak short-circuit currents, low operating losses and the correct operation of substation with interconnected busbars. The implemented system not only prevents economic losses due to the destruction or shortening of useful life of the equipment, but also increases the quality of energy supply to customers. It should be emphasised that this implementation project has been pioneered at a European level with the participation of large businesses related to the energy sector in addition to previous research and experimental tests performed in laboratories. The conclusion from the economic analysis is that the benefit-to-cost ratio of the technology would be positive with the estimated series-produced price for the system of 100 000 €.

Keywords: superconductor, fault current limiter, distribution substation

ABBREVIATIONS

AC	Alternating current
EDS	Supplying energy left
HTS	High temperature superconducting
HV	High voltage
MV	Medium voltage
SFCL	Superconducting fault current limiter
TIEPI	Time equivalent installed power interruption
YBCO	Yttrium oxide, barium and copper

INTRODUCTION

All the specialist literature agrees that in the medium- and long-term future, the energy sector will suffer major technological changes. The massive introduction of renewable energy, distributed generation and smart grids allows the growth of small electric power systems with islanding capabilities. The introduction of new technological equipment with more demanding electronic components, in terms of power quality, and advanced digital instrumentation, require a more demanding supply to consumers. This will be an excellent opportunity to further explore the weaknesses identified in legislation and standardisation, as well as the financial instruments needed to address the technology challenges. To avoid power outages, a system has been designed (a more detailed description will be provided later). The technology aims to provide electrical energy with no or practically no resistance (a property known as superconductivity) [1] with the consequence that electrical energy losses will be practically zero.

The system implemented is developed within the initiative of a European project, ECCOFLOW [2] and represents the latest research into superconductor technology applied to electrical networks [3]. Its main objective is to develop a current limiter based on second-generation high-temperature superconducting cables [4] (HTS) (Figure 1), all of which will now be abbreviated as SFCL. The ultimate goal is to integrate these systems into the substations of the electric distribution networks. The research on this advanced SFCL device operating on real distribution conditions is novel and complements the previous research.

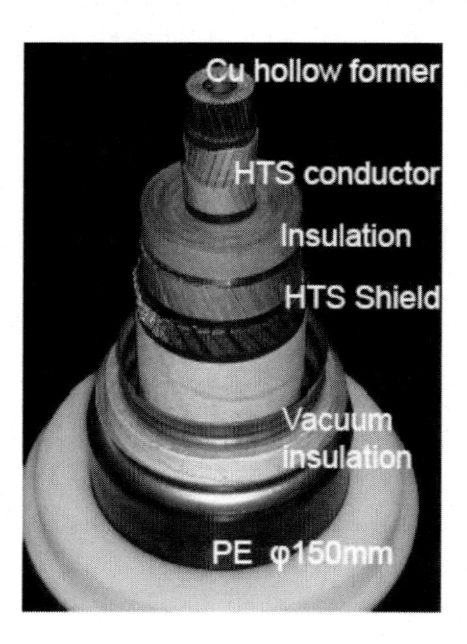

Figure 1. Cross-section of a high-temperature superconducting cable (HTS).

Figure 2. Location of the SFCL installation within the substation.

Through the installation of the SFCL, the aim is to improve the operation, stability, quality of the supply and efficiency [5] of the electrical networks without the need to construct additional substations or new infrastructure.

The project's main objective is to observe the behaviour and analyse the results in the medium- and long-term. Additionally, the system was adjusted in order to achieve the optimal operation. Besides, it is acceptable that this prototype, mainly due to its high initial investment cost, cannot have a short-term economic depreciation.

The ultimate goal of the SFCL system is none other than to ensure proper operation and achieve the objectives specified by the distribution company. The quality of supply is expected to increase, as it is affected by the various incidents [6] which may occur in medium-voltage networks and substations. Moreover, another consequence and result of the SFCL installation is that energy losses are minimised. Therefore, there are greater savings in costs for companies which have such a system installed.

It is important to note that, prior to its installation, a thorough analysis has been conducted to determine the ideal location for the installation of such a system with SFCL technology. It was an important part of the project, and researched together with the distribution company specialist staff. The primary objective of the analysis was the improvement of the quality of supply and the reduction of relevant negative incidents in HV/MV substations.

Before the installation process, the dimensions of the whole SFCL installation were taken into account. The feasibility was verified, since the minimum space required by the size of the SFCL installation is relatively small (below 100 m^2 as can be seen in Figure 2) with respect to the available land within the conventional medium-voltage substation in which it was installed.

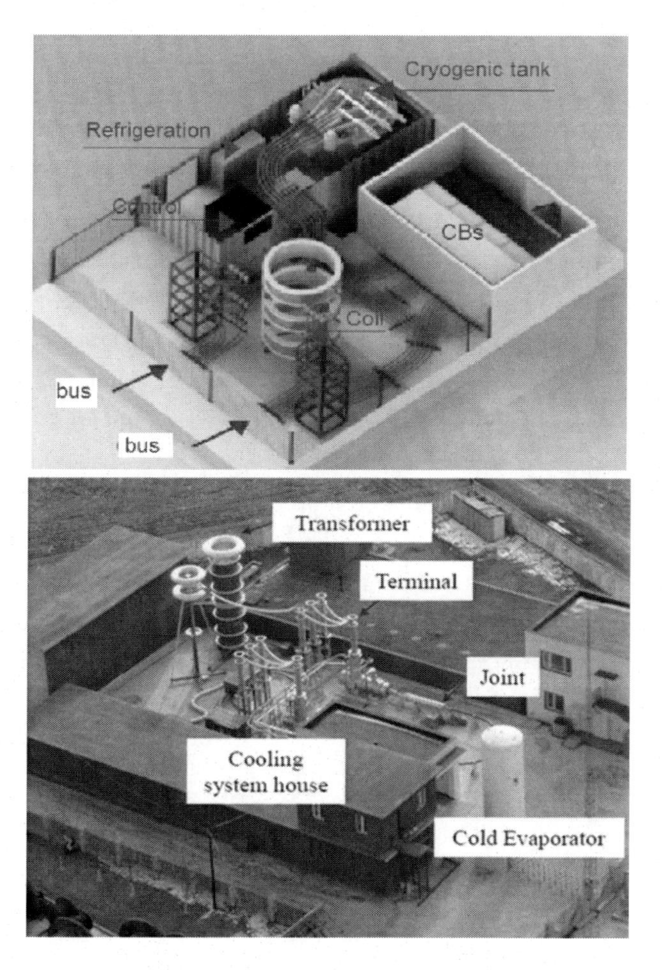

Figure 3. Design of the overall SFCL system installed.

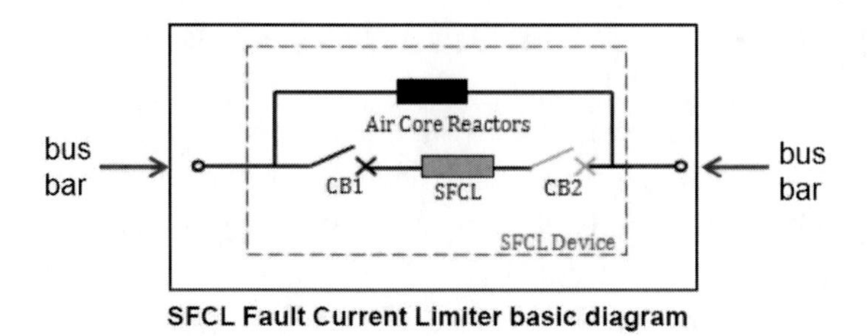

Figure 4. Basic diagram of the SFCL device installation.

The weight of the SFCL installation is also insignificant, as basic foundation and ground levelling are sufficient to support the weight of all components.

An overview of the design of the SFCL system is shown in Figure 3. This corresponds to the design process and the real installation. It shows all devices part of the installation, which will later be described in detail.

The SFCL in this project is arranged as medium-voltage longitudinal coupling busbars in a distribution network substation (Figure 4), which are in turn electrically fed by two processors working in parallel [7].

The descriptions and details of each of the components that make up the whole SFCL are found in section 2. Reference is also made to the original and final specifications required by the distribution company. Other relevant information about the components are also described therein. In section 3, the main advantages and disadvantages are listed together. Finally, in Section 4, the economic investment of the project and the results as an experimental prototype model for experimentation and analysis are stated and justified.

Description of the SFCL Installation Components

The various components of the SFCL installation have been manufactured [8] taking into account the electrical parameters for installation required and specified by the distribution company (Figure 5). These values have been supplied to the manufacturers to be taken into consideration when designing the different components so that the SFCL device can be seamlessly integrated into the preselected substation. However, this data is valid for any other SFCL device which is to be designed and installed within the same electrical distribution network, since variations in requests for electrical equipment [9] do not vary substantially from one substation to another.

rated voltage	16.5 kV
rated current	1000 A
max. prospective current (peak)	22 kA
max. limited current (peak)	10.8 kA
fault duration	1 s
HTS limitation time	80 ms
recovery time	< 30 s
AC withstand voltage	50 kV
lightning impulse	125 kV

Figure 5. Parameter specifications required by the distributing company.

Figure 6. Detailed drawing of the inside of the cryostat.

Figure 7. Design of the interior module of each deposit.

Firstly, the most important component in the SFCL assembly is the cryostat [10] (Figure 6). This element contains other devices which are detailed below. Since this project is a three-phase system, there are three tanks housed inside the cryostat, one for each phase, which in turn contain the superconducting material and one part of the cooling circuits.

Inside each of these deposits, a cylindrical module formed by concentric rings is housed (Figure 7). It comprises the high-temperature superconducting material (HTS), called so as it is a material that offers no resistance to current flow at a relatively high temperature with respect to other similar materials.

The conceptual design of the HTS module includes copper contacts which form the superconducting ring, as shown in Figure 8.

When designing the modules which comprise the superconducting material, the following adjustments and/or properties were taken into account:

1. Ability to withstand high intensity short-circuits to the superconducting material, recovering its initial state after a period of time from when the overload ceases.
2. Minimise current flow loss due to the distance between the bands which comprise the module being small.
3. Has been tried and tested for high voltages, above nominal value.
4. Reduced recovery time because the thermal interface is designed and optimised for most agents which can be used as refrigerants.
5. The design is optimised for subsequent mass production of the superconducting module.

It should be noted that depending on the material used as a superconductor, a certain coolant temperature will be required [11]. For materials made with copper compounds, the temperature should be maintained between 65K and 80K, in which the material acquires superconducting properties. In contrast, for other materials such as MgB2, the temperature range should be between 20K and 30K. The material chosen in this chapter was based on a YBCO composite [12], scientifically known as SF12100 (Figure 9). The refrigerant used is liquid nitrogen [13], because, amongst other reasons, it meets the specifications required and is one possible simple and cheap solution compared with other refrigerants. Its main advantages are its ease of marketing and acquisition in the market.

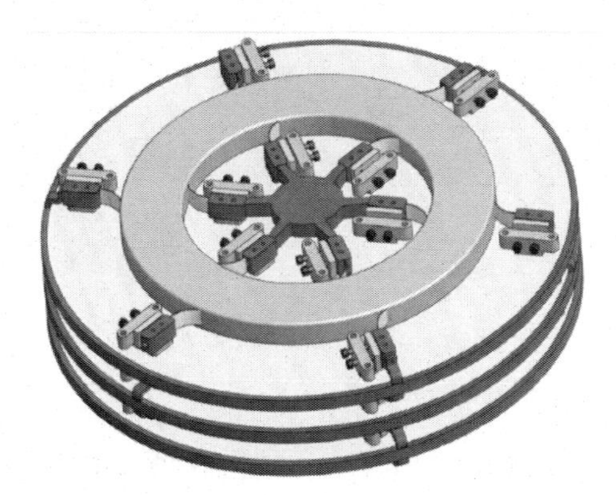

Figure 8. Details of each of the rings which form the inner tank module.

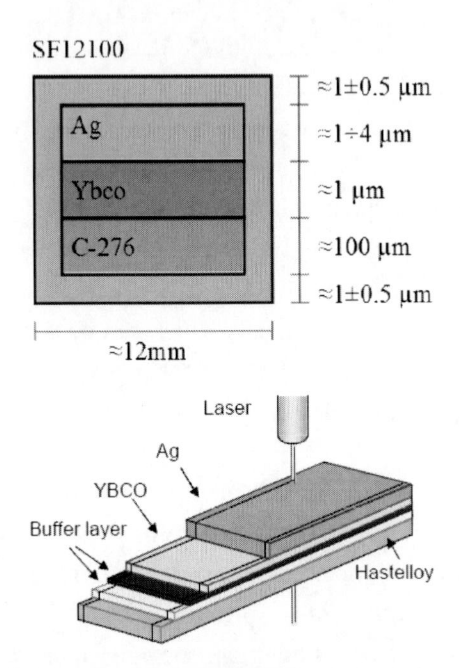

Figure 9. Composition and size of the superconducting material.

Figure 10. Container connections with the main circuit.

The connections between each of the containers which hold the superconducting material and the conductors forming part of the substation busbar circuit are shown in Figure 10. As can be seen, it is noteworthy that the location of the superconducting material is in series with the main circuit. Furthermore it includes shunt impedance [14] in parallel with the main circuit, whose mission is to limit the maximum current intensity passing through the SFCL. The value is easily configurable and, for this case, has been set by the distribution company considering the substation equipment installed in the SFCL device.

With respect to the refrigerant, this is driven by a set of compressor pumps through valves and circuits which ensure pressure, temperature and flow from the compressors located outside the cryostat to the inside in contact with the surface to be cooled. This whole system is called a cryorefrigerator [15] (Figure 11) and is an important and essential part of the SFCL assembly. As a property of the refrigeration and HTS material sources, the regenerative refrigerators [16] are the simplest option for this type of project. These provide localised cooling and therefore are viable for the chosen SFCL system.

With regard to the cryorefrigerator, we highlight the following elements:

- The current cost of the operation with the cryorefrigerator equals the price of liquid nitrogen for more than 10 kW of cooling power at 80K.
- For the above-mentioned refrigeration power range given in kW, the cost of liquid nitrogen in the market can be significantly higher. But any estimate of this prototype shows that the investment in the cryorefrigerator may only be recovered after several years of operation.
- The option of using liquid nitrogen as a refrigerant is even more attractive for refrigeration to 65K. This is because it decreases the need for increased performance for the cryorefrigerator. Since the liquid nitrogen undergoes partial vaporisation through isenthalpic expansion [17], it may even require the integration of the investment cycle in the circuit and compression pump.

The total consumption of the SFCL assembly for different ranges of current flow is detailed in Figure 12.

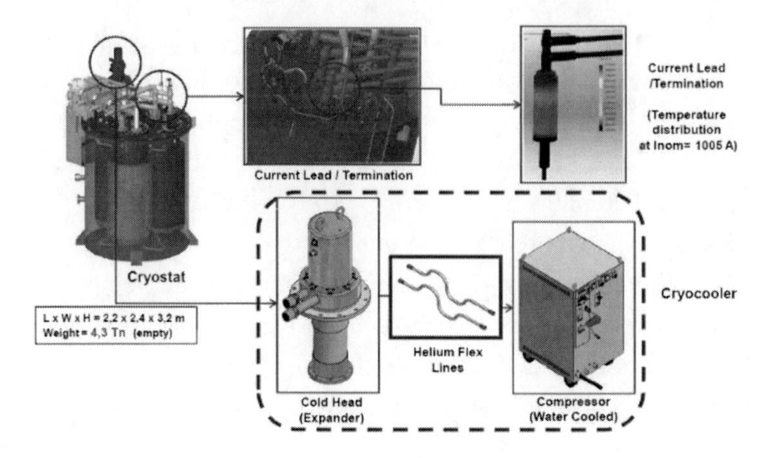

Figure 11. Detail of different parts of the cryorefrigerator.

Loss Contribution	Loss at 0,1 Ic	Loss at 0,5 Ic	Loss at 1 Ic
Max. superconductor AC loss 1)	< 1 W	≈ 10 W	150 W
Max. current lead loss 2)	180 W	≈ 220 W	270 W
Cryostat loss 3)	120 W	120 W	120 W
Max. additional loss 4)	1 W	15 W	60 W
Max. total loss at 77 K	≈ 300 W	≈ 365 W	600 W
Max. electric power at RT 5)	≈ 6.990 W	≈ 8.504 W	13.980 W
			8.000 W
Total Maximum Input Power			**≈ 22.600 W**

1) According to AC Loss report Ic = 300 A, L = 3,4 km
2) Specific current lead loss 45 W/kA
3) According to Cryostat Design
4) HTS-Copper-0.5Ω·12·2/3 = 4μΩ, Cooper connections - 2μΩ·12·2/3 = 16μΩ
5) GM Cryocooler efficiency (GM600) 1/23.3 and 6000 H MD data sheet

Figure 12. SFCL system consumption.

The simulation of the SFCL installation prototype was initially created adhering to few requirements which were determined by the distribution company (Figure 13). As the calculations were completed, the results for a first hypothesis were obtained (Figure 14) in which the first failure was produced in one of the medium-voltage lines [18]. In a second hypothesis (Figure 15), the failure occurs on the substation busbars [19]. In this latter case, it has been considered that there is no differential protection between busbars in the substation [20]. Therefore, in normal conditions, the failure would activate one of the overcurrent protections that exist (using nomenclature of the company 51) This corresponds to the opening of the switch on the low-voltage side of the transformer which feeds the medium-voltage busbars.

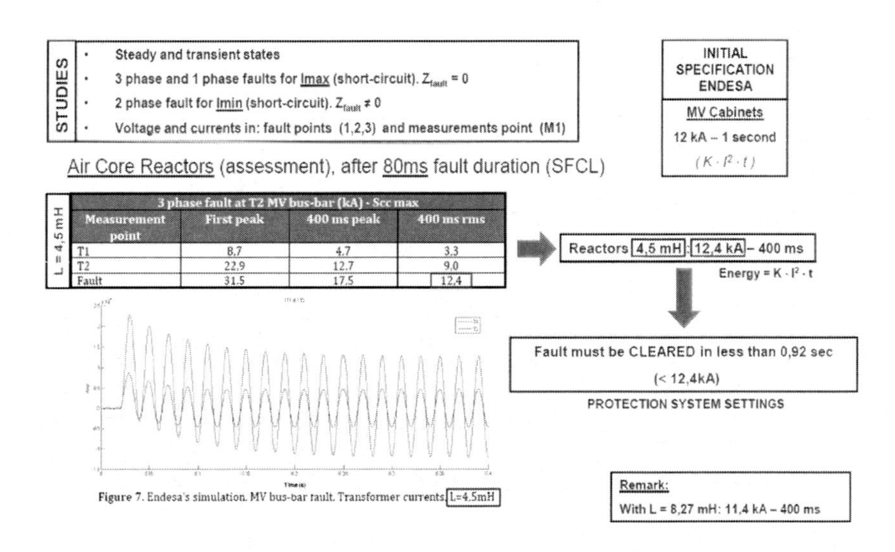

Figure 13. The distribution company's calculations and specifications.

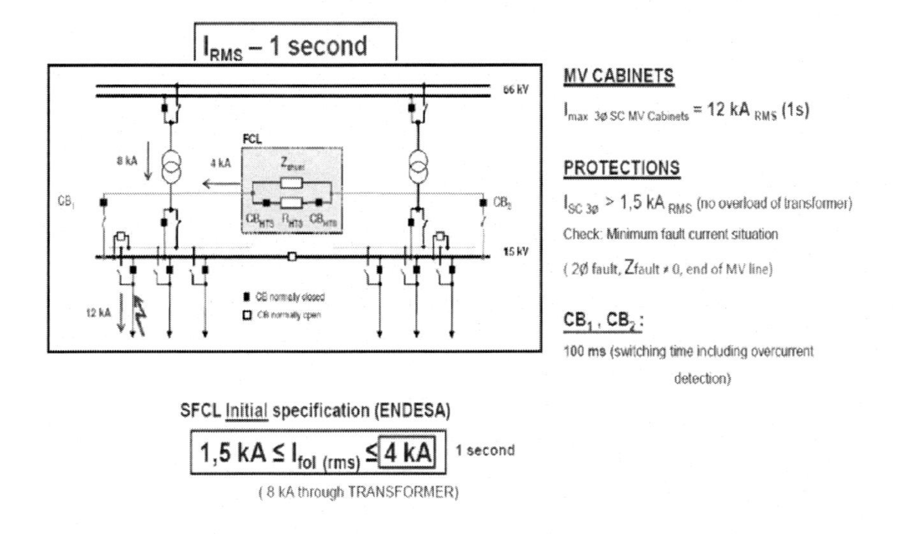

Figure 14. Calculation of currents for a simulated MV line failure.

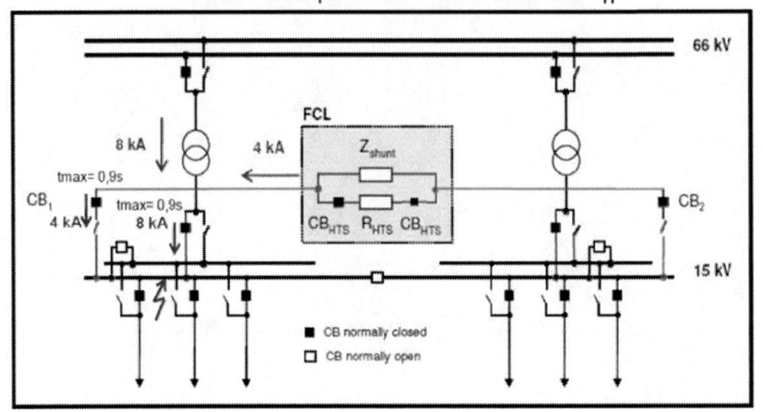

Figure 15. Calculation of currents for a simulated failure in the substation busbars.

After making several adjustments to the distribution network protection systems, the system specifications for the protection and control settings were again recalculated. The results, affecting only the first case described above, are shown in Figure 16.

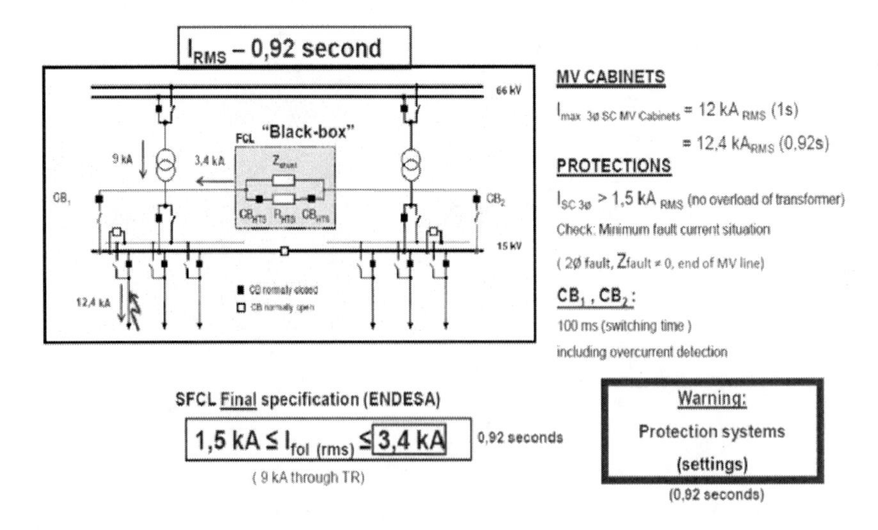

Figure 16. Final distribution company requirements.

The measurements which can be made and recorded on the SFCL during the beginning of the cooling process are detailed below:

- Changes in resistivity as a non-linear function of the current flow and temperature during cooling.
- Measurements of short circuit current values corresponding to the different impedances of the fault which occurs.
- Small currents due to the superconducting material including imperfections.
- Modelling and monitoring to find optimal stability.

Hotspots [21] are one of the greatest risk factors which may occur within the superconducting band inside the SFCL assembly, since that band as described above is not perfectly homogeneous. As a result of this consideration, unwanted effects which may be produced in the superconducting material must be considered and taken into account by the manufacturer. The mitigation for this risk is the correct quality control on the components, to avoid hotspots due to incorrect manufacturing. Other risks of using the SFCL are due to the greater complexity of the system. With increased number of components, the possibilities of failure are greater. This risk is mitigated with the bypass circuit to the equipment, although this bypass could also have problems with the additional breakers and connections. The values which may be reached in tests [22], carried out throughout the superconducting cable manufacturing process, must be specified. For obvious reasons, the greater the quality of the manufacturing equipment, the greater the associated cost.

One of the main advantages of SFCL modelling is that it is used to determine the safest and most stable form [23] of the electrical system. Therefore, it can be used to draw a series of conclusions. The most important and relevant of which is that there are no problems with low impedance. The most critical calculated and tested problems for the SFCL assembly in the project occur at around 15 Ω of impedance for the specifications given in Figure 5. The stabilisation of the whole system is also able to support the most critical events [24].

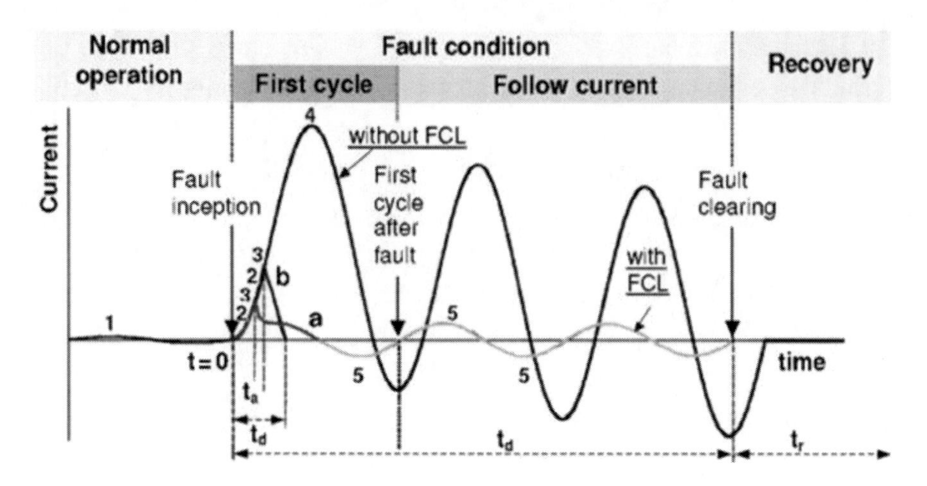

Figure 17. Behaviour of current waveforms during a failure with installed SFCL.

where:
U_n: rated system voltage
t_a: action time (from t = 0 until I_{max})
t_d: fault duration time
t_r: recovery time
time:

1 ➔ rated current (I_n)

2 ➔ minimum initiating current (I_{min})

3 ➔ maximum limited current (I_{max})

4 ➔ peak (prospective) short circuit current (I_p)

5 ➔ peak value of the follow current (I_{fol})

The operation of this whole SFCL system, as described above, is integrated into the busbars of a conventional medium-voltage substation. The short-circuit current which runs through it can be reduced from the first cycle [25] like a variable resistor. During normal operating range, when the current flowing through the superconductor material is within the operating range and shows no short circuit currents or imbalances, the superconductor material providing almost no resistance (Figure 17).

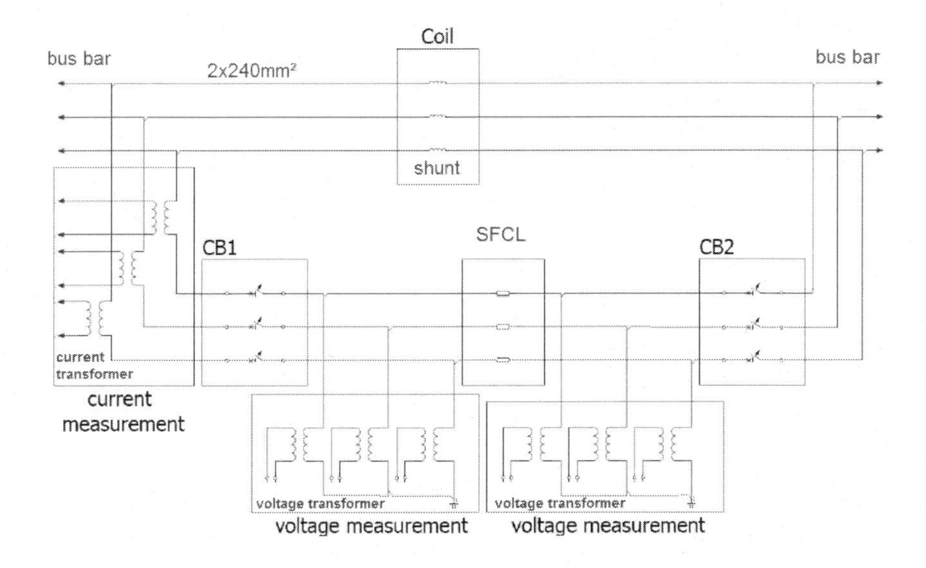

Figure 18. SFCL wiring diagram between medium-voltage busbar switches.

On the SFCL diagram shown in Figure 18, the single-line diagram can be seen, including an input (CB1) and an output switch (CB2) for disconnection of the whole grid.

The electrical connection of the SFCL components to the distribution network is detailed in Figure 18, where the two three-phase differential circuits which are connected in parallel can be seen:

- The lower (parallel) circuit, where the switches CB1 and CB2 are located has the main function of isolating the main superconducting circuit (marked as SFCL). It includes a series of protective measures [26], taking measurements of their respective current transformers (to measure current) and voltage transformers (to measure voltage). These protections are also set, taking into account the criteria and parameters set by the power distributing company according to their needs [27]. Firstly, using the current measured at the entrance of the main circuit and, secondly, the measured voltages at the input and output respectively of the same circuit.

- In the upper circuit (secondary), shunt impedance (inductive) is connected in series with the circuit. The basic function is to limit the current in the circuit in case of disconnection of the primary circuit at a nominal value. Consequently, to provide electrical continuity to both ends of the circuit where the SFCL assembly is connected.

With this approach, the idea is to provide all necessary information to adequately assess the quality of the supply and the energy losses. These have to be compared with other possibilities associated with guaranteeing the ideal clearance conditions. Therefore, the development of this prototype is framed in the context defined above. During the preparation, the main proposed protection criteria have been provided. The SFCL model which is installed may undergo subsequent revisions, based on the experience gained from its implementation.

ADVANTAGES AND DISADVANTAGES OF THE SFCL SYSTEM

The main advantages (Figure 19) of the installed system are represented in Figure 19.

- The main economic advantage is avoiding having to invest in renovation of conventional facilities to meet and comply with the provisions and enforcement of current legislation.
- The main technical conditions, such as the quality of the supply and transformer yield losses mainly justified by:
  - The high voltage electrical networks "upstream" of the SFCL installation should not be affected.
  - Better load balancing with transformers working in parallel [28], since the power demand from the medium-voltage busbars is shared equally among all substation transformers.

This results in better and more efficient use with the consequence of the optimisation of the installed transformer capacity and of minimising energy losses (Figure 20)

- As a result of this it is possible that, by sharing the demand, there is no need to expand the facility by adding another transformer (Figure 21).
- Instead of disconnecting one of the transformers from the network, the other transformer may continue to feed the total energy demand of the medium-voltage busbars to which they are connected (known as N-1 criteria) [29], so there is no supply disruption.
- Increase in the short-circuit power due to the configuration of the transformers [30], therefore transferring more energy to the distribution network through special system providers.

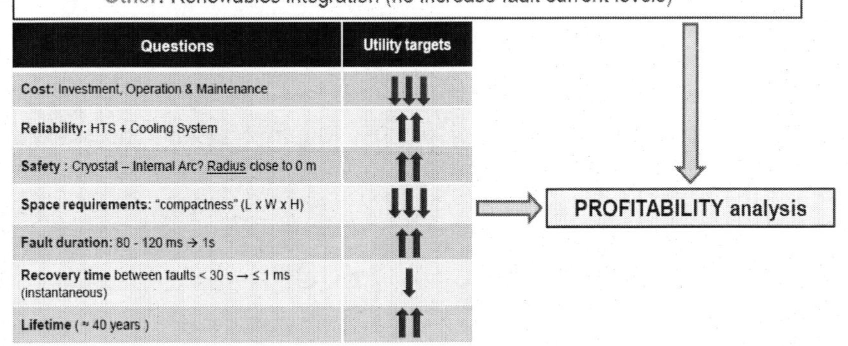

Figure 19. Advantages of the SFCL installation.

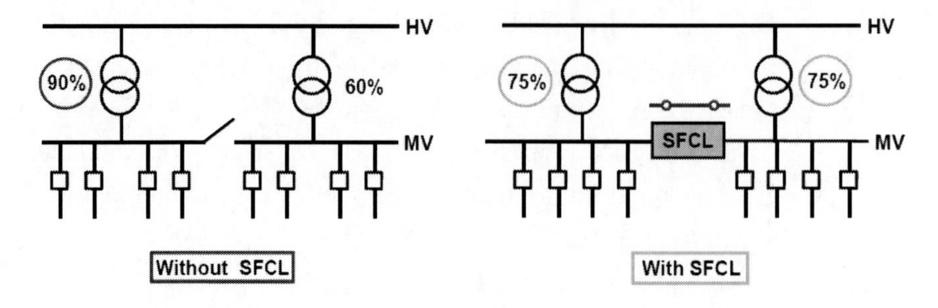

Figure 20. Load balancing and minimising energy losses with SFCL.

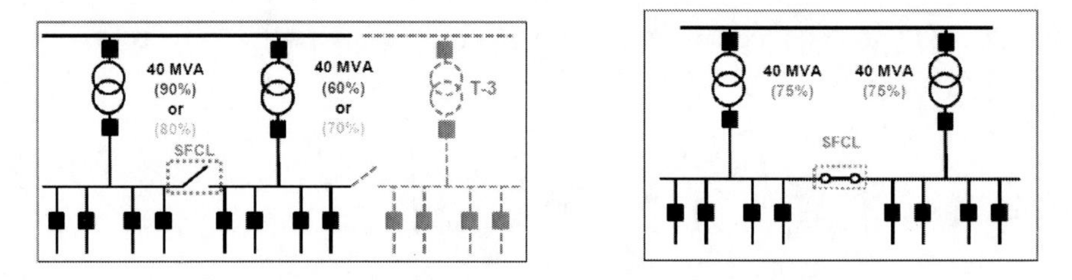

Figure 21. Load distribution avoiding the need for another transformer.

- Other criteria, such as the integration of more renewable energy. Increasing the short circuit power if parallel buses can be connected allows more interconnection of renewable power. Spanish legislation provides a power limitation to the connection of renewable energy sources (photovoltaic, wind power, etc.) to the MV network, setting the limit of 5% on the maximum short-circuit power at the point of connection. Installing a SFCL in a HV/MV substation by applying the coupling busbars increases the short circuit power due to the parallel connection of the substation transformers. This increase in power affects any point "downstream" connected to the MV substation, resulting in a potential increase in renewable energy sources connected to it.

Regarding the disadvantages, these can basically be summarised into three:

- Firstly, the high cost of acquisition and implementation of the SFCL at the present moment.

- Secondly, the difficulty of finding qualified personnel skilled in the subject. Knowledge and time is required from the preliminary analysis stage until implementation, execution and commissioning of the SFCL. Finally, also for the data acquisition, its analysis and modelling.
- Thirdly, the operation and maintenance of the system, requires specific knowledge and is different to normal substation maintenance. The behaviour after several years of operation will have to be assessed.

ECONOMIC JUSTIFICATION

The implementation of the SFCL assumes the high cost of the entire assembly, due mainly to its design as a prototype being unique. The system was specifically designed to test, as a pioneer installation, the behaviour in limiting failures in a substation coupling busbars. The SFCL purchase price also includes the costs of the trials and tests in different laboratories before installation. This justifies that the resulting cost of the SFCL assembly was over one million euros, roughly 1.030.000 €. However, as has already been mentioned and justified above, the main purpose for the implementation of the SFCL device is analysing its behaviour in real situations, studying the results obtained, drawing conclusions and being able to model future devices to be manufactured.

However, the benefits presented in section 3 can be economically estimated in order to calculate the profitability of the SFCL.

AVOIDED COST FOR NON-DELIVERED ENERGY

In this case it has been estimated that the value for energy that will be delivered is 25430 kWh. The value has been estimated based on the expected faults affecting the substation. Given the mean power delivered is

36 MW, the expected interruption time due to short-circuit causing trips is a total of 42 minutes per year. This value is for the existing substation and network conditions, and is also an estimation based on operational experience.

Hence, it is the energy that would not be delivered in the absence of the SFCL system because of interruptions. In order to calculate the avoided cost the price of the rates established for the energy term of 0.149198 €/kWh (fixed tariff) set in Spain is used. The result is 3794 €/year of increased value.

INCREASE EFFICIENCY FOR PARALLEL TRANSFORMER OPERATION

The SFCL, by reducing the short circuit current in case of fault, allows parallel operation of the transformers in the substation. The energy cost savings is thus by the reduction of the losses in copper, as the point. According to the simulator program of the distribution company the result of economic losses in this substation has been for 2014 of 605 €/year. The reason behind is a lower loading of the transformers, while the reduced load losses outweighs the no-load losses in the system.

This increase in efficiency has to be compared with the additional losses in the SFCL, as depicted in Figure 12 and is mentioned below.

AVOIDED COSTS OF TRANSFORMER DISCONNECTION

In the event of a trip of one of the two transformers, considering that the substation meets the criteria N-1, the other transformer is able to take full demand without affecting the market. While in 2014 there was no trip event on one of the transformers, the possibility should never be ruled out. Remuneration of distribution is influenced by TIEPI values. Thus, considering the probability of the event and the penalization cost is estimated at 6.542 €/year.

Avoided Costs for Generator Unavailability

There are losses according to the incidents referred to under Section 3 of this study, producing the unavailability of the generation group/s. These can be avoided with the use of the SFCL system. These economic losses are quantified as directly proportional to the downtime, the higher the downtime of the generator, the greater the economic loss. The estimated average value is set as 9,887 €/year for the specific site conditions and affected generators.

Avoided Costs for Replacement, Repair and Maintenance

The incidents mentioned in the previous section not only cause unavailability of groups, but sometimes are likely to lead to a malfunction or other problems in the components of the generators. Avoided costs for repair, replacement, maintenance, include spare parts or components affected and labor involved. The average for the considered incidents is 12.326 €/year.

Avoided Cost for Equipment Life Time

When short circuits occur, the operational life of electrical or electronic devices is reduced. Some of the components substantially reduce its life, for example, the circuit breakers with the number of operations. Manufacturers of the equipment calculate the life time for a number of operations. The reduction of switching operations by the use of a SFCL device avoids the replacement of equipment, extending the operational life. A similar life reduction happens to cables, where short circuits reduce the lifetime exponentially. The calculation of depreciation of the assets in this case, caused by the shortening of the life thereof is estimated as 759 €/year.

The resulting benefit of avoided costs for the SFCL installation is 33.913 €.

As was mentioned previously, considering the cost of the prototype (1.030.000 €), the system would not be profitable. The pay-back time will be close to 30 years, above the operational life of the system. Nevertheless, the price of the SFCL to be considered is from serial production, not the prototype cost.

Considering the data provided by the different suppliers, it is determined that the cost of the SFCL, installed and tested, once the series production is reached, will be 100.000 €. The system does not require yearly maintenance costs, only the SFCL remote monitoring and associated alarms, and the substation external maintenance, performed regularly by the substation maintenance personnel. The cost of the losses of the system is small, but has to be calculated to compare with the efficiency gains. With a steady-state 150 W consumption, the yearly cost is 195 €. Resulting yearly benefit is therefore 33.718 €. Taking into account the losses at 1 p.u. could be maximum 600 W, the maximum cost would be 780 €. If the losses are at this level, for 8760 h, the benefit of parallel operation of transformers would be lost with the losses in the system and the cryogenic cooling.

Considering these expected annual savings, based on statistical data and simulations of the Distribution System Operator, the SFCL proves profitable. The investment in a series produced SFCL could have an estimated payback time of 3 years.

CONCLUSION

During the research project, the study of a novel SFCL installed in a power distribution substation has been completed. The design and implementation of a SFCL system, is described in this chapter, and is a unique prototype for the researching, experimentation and development of this technology. In addition, the system represents an opportunity for learning and trialling the actual behaviour of all components of the SFCL assembly. All of the parties involved in the development of the system presented in the chapter include the European Development Fund and ten private international electrical companies. The results have been positive, achieving the expected performance of the SFCL assembly, with reduced losses, short circuit limitation, and allowing interconnected busbar operation of the substation.

As one of its immediate and practical applications, this type of advanced SFCL technology could be necessary for interconnected networks or smart grids (Figure 22). The automated fault limitation is a functionality that would increase the lifetime of the equipment and increase reliability, without compromising energy efficiency.

The SFCL system is expected to be used more widely with the estimated series-production that could bring the system cost to 100.000 €. In this scenario the benefits would outweigh the costs and increased installation of these systems is expected. Generally, the economic profitability of the system will depend on the site conditions and the specific costs avoided using an SFCL system.

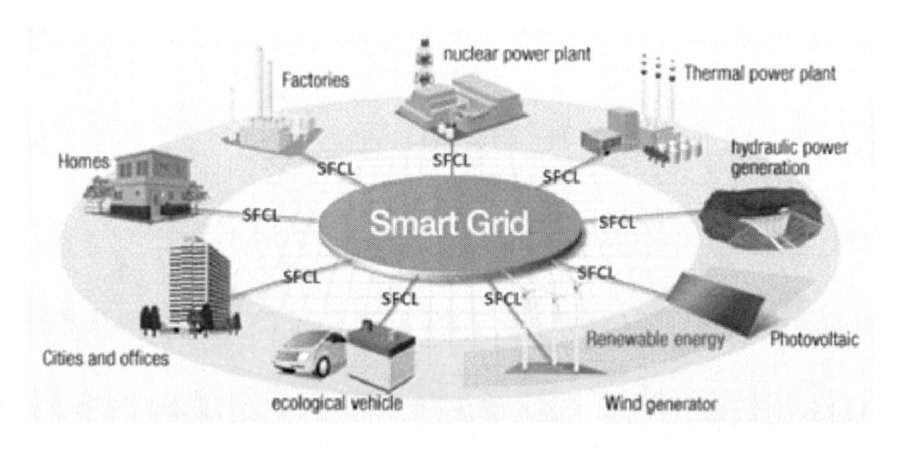

Figure 22. SFCL integrated in smart grids.

References

[1] Charles P. Poole Jr.,. Farach, Horacio A., Creswick, Richard J., Prozorov, Ruslan. BCS Theory. *Superconductivity* (Second Edition), 2007, Pages 171-193.

[2] Eccoflow. "Development and field testing of an efficient YBCO Coated Conductor based Fault Current Limiter for Operation in Electricity Networks," *Eccoflow.* http://www.eccoflow.org/ (June 2014).

[3] Noe, M., Tixador, Hobl, P., Martini, L., Dutoit, B. "Conceptual Design of a 24 kV, 1 kA Resistive Superconducting Fault Current Limiter," *IEEE Transaction on Applied Superconductivity*, 2012, vol. 22, article 5600304.

[4] Hemdan, Nasser G. A., Kurrat, Michael, Schmedes, Tanja, Voigt, Antje, Gandioli, Rüdiger Busch C., Tixador, P., Mariani, G. Bueno. "Integration of superconducting cables in distribution networks with high penetration of renewable energy resources: Techno-economic analysis," *International Journal of Electrical Power & Energy Systems,* Volume 62, November 2014, Pages 45-58.

[5] Moreno-Muñoz, A., de la Rosa, Juan José González, Flores-Arias, J. M. F., Bellido-Outerino, J., Gil-de-Castro, A. "Energy efficiency criteria in uninterruptible power supply selection," *Applied Energy,* Volume 88, Issue 4, April 2011, Pages 1312-1321.

[6] Bennett, Simon. "Insecurity in the Supply of Electrical Energy: An Emerging Threat?," *The Electricity Journal,* Volume 24, Issue 10, December 2011, Pages 51-69.

[7] Borge-Diez, David, Colmenar-Santos, Antonio, Castro-Gil, Manuel, Carpio-Ibáñez, José. "Parallel distribution transformer loss reductions: A proposed method and experimental validation," *International Journal of Electrical Power & Energy Systems*, Volume 49, July 2013, Pages 170-180.

[8] Teng, Yuping, Dai, Shaotao, Song, Naihao, Zhang, Jingye, Gao, Zhiyuan, Zhu, Zhiqin, Zhou, Weiwei, Wei, Zhourong, Lin, Liangzhen, Xiao, Liye. "Analysis on heat loss characteristics of a 10

kV HTS power substation," *Cryogenics,* In Press, Corrected Proof, Available online 24 March 2014.

[9] Bedkowski, Mateusz, Smolka, Jacek, Banasiak, Krzysztof, Bulinski, Zbigniew, Nowak, Andrzej J., Tomanek, Tomasz, Wajda, Adam. "Coupled numerical modelling of power loss generation in busbar system of low-voltage switchgear," *International Journal of Thermal Sciences*, Volume 82, August 2014, Pages 122-129.

[10] Young, E. A., Bailey, W. O. S., Al-Mosawi, M. K., Beduz, C., Yang, Y., Chappell, S., Twin, A. "A Dual Operational Refrigerator/Flow Cryostat with Wide Bore Medium Field Magnet for Application", *Physics Procedia*, Volume 36, 2012, Pages 1343-1347.

[11] Lewandowska, Monika, Wesche, Rainer. "Parametric study for the cooling of high temperature superconductor (HTS) current leads," *Cryogenics,* Volume 53, January 2013, Pages 31-36.

[12] Gandioli, C., Tixador, P., Bueno Mariani, G. "Test and simulations of different YBCO tapes for FCL," *IEEE transactions on Applied Superconductivity*, vol. 22, 2012, Issue 3, Article 5603104.

[13] Watanabe, M., Yumura, H., Hirota, H., Masuda, T., Shimoda, M., Ohno, R., Ikeuchi, M., Yaguchi, H., Ichikawa, H., Mimura, T., Honjo, S., Hara, T. "Recent Progress of Liquid Nitrogen Cooling System (LINCS) for Yokohama HTS Cable Project," *Physics Procedia*, Volume 36, 2012, Pages 1313-1318.

[14] Torres, V., Guardado, J. L., Ruiz, H. F., Maximov, S. "Modeling and detection of high impedance faults," *International Journal of Electrical Power & Energy Systems*, Volume 61, October 2014, Pages 163-172.

[15] Yamada, K. "Development of a large cooling capacity single stage GM cryocooler," *Cryogenics,* In Press, Corrected Proof, Available online 13 April 2014.

[16] Paurine, A., Maidment, G. G., Eames, I. W. "Development of a packed bed regenerative solution heat exchanger (R-SHX) for a single stage LiBr–H2O vapour absorption refrigeration (VAR) system," *Applied Thermal Engineering*, Volume 60, Issues 1–2, 2 October 2013, Pages 182-187.

[17] Lee, Hoseong, Hwang, Yunho, Radermacher, Reinhard, Chun, Ho-Hwan. "Potential benefits of saturation cycle with two-phase refrigerant injection," *Applied Thermal Engineering*, Volume 56, Issues 1–2, July 2013, Pages 27-37.

[18] Didier, G., Lévêque, J. "Influence of fault type on the optimal location of superconducting fault current limiter in electrical power grid," *International Journal of Electrical Power & Energy Systems*, Volume 56, March 2014, Pages 279-285.

[19] González, A., Echavarren, F. M., Rouco, L., Gómez, T., Cabetas, J. "Reconfiguration of large-scale distribution networks for planning studies," *International Journal of Electrical Power & Energy Systems,* Volume 37, Issue 1, May 2012, Pages 86-94.

[20] Noshad, Bahram, Razaz, Morteza, Seifossadat, Seyed Ghodratollah. "A new algorithm based on Clarke's Transform and Discrete Wavelet Transform for the differential protection of three-phase power transformers considering the ultra-saturation phenomenon," *Electric Power Systems Research*, Volume 110, May 2014, Pages 9-24.

[21] Jadin, Mohd Shawal, Taib, Soib. "Recent progress in diagnosing the reliability of electrical equipment by using infrared thermography," *Infrared Physics & Technology*, Volume 55, Issue 4, July 2012, Pages 236-245.

[22] Martini, Luciano, Bocchi, Marco, Ascade, Massimo, Valzasina, Angelo, Rossi, Valerio, Angeli, Giuliano, Ravetta, Cesare. "Development, Testing and Installation of a Superconducting Fault Current Limiter for Medium Voltage Distribution Networks," *Physics Procedia,* Volume 36, 2012, Pages 914-920.

[23] Kumar, Deepak, Samantaray, S. R. "Design of an advanced electric power distribution systems using seeker optimization algorithm," *International Journal of Electrical Power & Energy Systems*, Volume 63, December 2014, Pages 196-217.

[24] Garcez, Thalles Vitelli, De Almeida, Adiel Teixeira. "A risk measurement tool for an underground electricity distribution system considering the consequences and uncertainties of manhole events,"

Reliability Engineering & System Safety, Volume 124, April 2014, Pages 68-80.

[25] Ouyang, Jinxin, Xiong, Xiaofu. "Research on short-circuit current of doubly fed induction generator under non-deep voltage drop," *Electric Power Systems Research,* Volume 107, February 2014, Pages 158-166.

[26] Chen, L., Tang, Y. J., Shi, J., Ren, L., Song, M., Cheng, S. J., Hu, Y., Chen, X. S. "Effects of a voltage compensation type active superconducting fault current limiter on distance relay protection," *Physica C: Superconductivity*, Volume 470, Issue 20, 1 November 2010, Pages 1662-1665.

[27] Filipović-Grčić, Dalibor, Filipović-Grčić, Božidar, Capuder, Kosjenka. "Modeling of three-phase autotransformer for short-circuit studies," *International Journal of Electrical Power & Energy Systems*, Volume 56, March 2014, Pages 228-234.

[28] Wang, Dan, Mao, Chengxiong, Lu, Jiming, He, Jinping, Liu, Haibo. "Auto-balancing transformer based on power electronics," *Electric Power Systems Research,* Volume 80, Issue 1, January 2010, Pages 28-36.

[29] Luo, Fengzhang, Wang, Chengshan, Xiao, Jun, Ge, Shaoyun. "Rapid evaluation method for power supply capability of urban distribution system based on N − 1 contingency analysis of main-transformers," *International Journal of Electrical Power & Energy Systems*, Volume 32, Issue 10, December 2010, Pages 1063-1068.

[30] Lei, Xiao, Li, Jian, Wang, Youyuan, Mi, Sibei, Xiang, Chengmeng. "Simulative and experimental investigation of transfer function of inter-turn faults in transformer windings," *Electric Power Systems Research,* Volume 107, February 2014, Pages 1-8.

In: Renewable Electric Power Distribution … ISBN: 978-1-53614-202-0
Editors: A. Colmenar-Santos et al.

Chapter 3

DISTRIBUTION TRANSFORMER LOSS REDUCTIONS

África López-Rey[1], Manuel Castro-Gil[1,*] and Carlos-Ignacio Cuviella-Suárez[2]

[1]Departamento de Ingeniería Eléctrica, Electrónica, Control, Telemática y Química Aplicada a la Ingeniería, Universidad Nacional de Educación a Distancia (UNED), Madrid, Spain

[2]ROCA, Barcelona, Spain

ABSTRACT

Transformers in electrical distribution systems for buildings and industries are set up to ensure the continuity of supply. Modern distribution transformers are reaching increasingly higher levels of energy efficiency but contribute to between 16% and 40% of the energy losses associated with electrical distribution systems. This chapter examines and proposes a method to reduce losses in transformation systems and thus to

* Corresponding Author Email: mcastro@ieec.uned.es.

reduce the associated emission of greenhouse gases (GHGs). The method has been validated in 12 real facilities of different power and efficiency levels. Reductions in the losses of the studied transformers were up to 41% with respect to the initial losses. This demonstrates the beneficial operation of the PLO method proposed in this chapter in a wide range of existing transformers or for future installations. The research has obtained patent pending status P201101267 in Spain.

Keywords: energy efficiency, distribution transformer, greenhouse gas emissions, energy conservation, energy systems analyses, distribution system

NOMENCLATURE

DEPT	Distribution Electronic Power Transformer
GHG	Greenhouse Gas
L	Load (kVA)
PLC	Power Line Communication
PLO	Parallel Losses Optimization (kVA)
PL	Percentage Losses
S	Transformer Rated Power (kVA)
TOC	Total Owner Cost ($/kW yr)
U	Transformer percentage impedance (%)

Subscripts

0	No load
1	Parallel transformer 1
2	Parallel transformer 2
dem	Demanded
L	Losses (kVA)
N	Nominal
opt	Optimal
par	Parallel
SC	Short Circuit
T	Total

INTRODUCTION

Transformation systems installed in buildings and industries connected to distribution networks ideally should ensure continuity of the energy supply at the point of consumption in the most efficient manner possible. Different international regulations require the assurance of supply through the installation of two or more parallel power transformers, such that maintenance operations can be performed on one of transformer without cutting off the supply; additionally, such an arrangement guarantees continuity of operation in case of failure. The total losses in a transformer are the sum of the no load and load losses. Over time, the energy efficiency of distribution transformers has increased, and the total losses have been reduced, representing an important advancement in electrical systems.

Transformation systems are an important percentage of the losses in electrical distribution systems, both in Europe and United States [1]. Despite the fact that transformer manufacturers [2] have systematically reduced losses by introducing new materials and manufacturing techniques [3-7], a significant amount of already installed equipment possesses energy efficiency values that are significantly lower in comparison to the systems manufactured today. Modern transformer manufacturing techniques allow development of systems with lower losses in comparison to the traditional systems, further involving reductions in the associated greenhouse gas emissions [5, 6]. Transformer systems with low-losses reduce the instantaneous power demanded by the installation and, therefore, reduce the power associated with generation. In the case of electricity generation based on fossil fuels, this implies a reduction of CO_2 equivalent on the order of tons. With respect to systems installed in distributed generation grids [8, 9], using renewable generation implies a reduction in the installed power and facilitates a better forecast of demand and enhanced system control.

The necessary investment in power distribution systems and the cost of electricity generation continue to increase; hence, technologies to reduce energy consumption are significantly in demand at present. Both electric power suppliers and final consumers in private facilities are directly benefited by the implementation of systems that reduce losses in the distribution transformers. Transformers are operated in every power distribution system, so that the energy savings associated with new technologies or transformation systems that increase the efficiency reach high values. Additionally, the newly installed transformers have no associated load losses, preventing secondary energy consumption without demand. In the design phase for a new installation, the choice between the use of a system of high efficiency versus one with major losses is directly related to the expected cost savings during use of the equipment. Analysis of the Total Owning Cost (TOC) is used as a tool for decision making by taking into account the sum of the cost of the transformer to own and the costs arising from losses during its lifetime [10]. Examples of the implementation of this method for the evaluation of facilities [11, 12] have been reported, depending on the type of use [13-15] and specific to industrial and commercial facilities [16, 17]. The TOC calculation evaluates losses in the transformer for load and no load modes to determine which type of transformer is the most appropriate based on the cost of losses during the life cycle of the equipment. This tool is used to make decisions in the design phase [18, 19], specifically for manufacturing and purchasing equipment [20]. In this chapter, the TOC is analyzed, and a new method is proposed, called PLO, which is aimed at the minimization of losses in the system during the device's entire life cycle.

For existing facilities with transformers in operation, it is necessary to consider techniques that allow reductions in the energy losses associated with transformers without requiring replacement of the existing transformers. The replacement of a device that is not damaged is justifiable if, after using a TOC analysis, the result profitability advice indicates discarding the existing system and the installation of a new transformer. This substitution is not economically viable in almost any case, so it is necessary to establish energy saving alternatives. For new facilities, the choice to use higher efficiency transformers can be profitable when the reduction in energy losses justifies the cost increase in comparison to that of a conventional transformer. Further, the existing transformers must be integrated in the new distributed generation systems, and the proposed approach for loss reduction methods facilitates this task, reducing the instantaneous power demand in the grid. In industrial facilities and commercial centers, transformers usually remain connected regardless of the power demand. However, many facilities exist with hourly, daily and monthly seasonality of use, so the transformation capacity installed is not always optimally adjusted. Additionally, based on forecasted expansions in the installation over time, it is typical to oversize transformer systems by a minimum value of 20%. Generally, the selection of one type of transformer with a large rating gives the maximum efficiency, which is further associated with simpler installation in comparison to use of more than type of one transformer. In large plants, two or more transformers of equal rating may be selected. This mode of operation is frequently required, but to ensure adequate performance, both transformer types must possess the same voltage ratio, the same percentage impedance, the same polarity and the same winding connections. For critical continuous operation plants or facilities, e.g., hospitals, power may be provided by two independent feeders. The feeders can be at similar or different voltage levels. In all such cases, each transformer may be capable of running the plant, such that in its normal operation each transformer only experiences 50% load. For non-continuous demand facilities, the load may be lower than 25% at times. For the non-continuous operation of plants with holidays or seasonal industries, switching one transformer to an off mode to save a portion of the load losses is generally considered, although the procedure has not been automated. The method proposed in this chapter allows the transformers power to be adapted to the demanded power, incurring possible minor losses by connecting the transformer that ensures the required power and reducing the overall energy losses. Additionally, the proposed PLO method allows reduced losses in all types of facilities using parallel transformers, which is applicable to both existing and new systems, thereby allowing energy savings regardless of the efficiency of the transformer class.

This chapter investigates and proposes a method of optimization for transformer utilities that is based on use of two transformers in parallel, as a standard solution for both industrial and commercial entities. This parallel approach is analyzed based on the losses and the reduction in losses associated with implementation of this method. A wide range of transformer combinations, with powers ranging from 100 kVA to 1,600 kVA, were analyzed, providing transform powers from 200 kVA to 3,200 kVA. Three transformer classifications, offering high, medium and low losses, were analyzed for each rated power. The transformers were characterized according to the parameters of load losses and no load losses, and the parallel behavior of the transformer sets was studied for all points of operation. The points of operation were calculated, and the total losses in the transformation utility for one or the other or both connected transformers were characterized. The Parallel Losses Optimization (PLO) was defined as the point of rupture; that is, the point at which it is more profitable to have connected only one transformer or both in parallel was established for each transformation power investigated. Additionally, the demand curves of four installations with installed power ratings of 650 kVA, 1,260 kVA, 1,630 kVA and 2,600 kVA were studied to experimentally determine the potential savings. Several companies in the sector of transformer maintenance in Spain were consulted to shed light on the maintenance protocols for the systems and to calculate the potential savings associated with operation of the transformation system according to the PLO proposed in this chapter. Transformer systems are widely used worldwide and will continue being installed during the coming years; hence, the energy saving potential is significantly high. The use of distributed generation systems will also increase the number of installed transformers.

The chapter is organized as follows. Section 2 discusses the losses associated with transformers, both in parallel and unitary operation. Section 3 describes the methodology of the study carried out; in section 4, the results are evaluated. The conclusions are presented in section 5. Due to the general interest associated with this invention and the opportunities for cost reduction in electrical systems, the approach presented here is currently under patent pending status.

ENERGY LOSSES IN TRANSFORMERS

Transformer Losses

Transformers are electric machines with high performance (more than 95%). The transformer is based on the use of two or more windings around a ferromagnetic core. The tension level changes without a change in frequency. Transformers are widely used in electrical distribution systems to carry out basic functions because they act as isolation elements and voltage level transformers. Due to the use of copper and ferromagnetic materials their cost is appreciable, and they are heavy and bulky pieces of equipment. Reducing the size of these machines depends on the density of power in the transformer, which is inversely proportional to the frequency; hence, systems that work at high frequencies allow more efficient use of the magnetic core with associated reductions in size [21, 22]. Currently, multiple research efforts on power electronics-transformers are underway. These systems, called Distribution Electronic Power Transformers (DEPTs), allow the quality problems associated with energy distribution systems to be addressed more effectively in addition to reductions in losses [23-29]. DEPT systems are under development but have not yet been introduced in electrical distribution systems.

This chapter focuses on the ferromagnetic systems currently installed in distribution systems, specifically those that will remain installed over the next several years. According to studies carried out in the United States, distribution transformer losses account for 40% of the losses in non-generating public utilities and 16% in investor-owned private utilities [1]. Due to the high cost of the new systems, replacing distribution transformers cannot be carried out in an extensive manner; hence, in this chapter, a method for reducing the losses in distribution transformers regardless of their efficiency level is presented [30-33]. In June, 2011, the International Energy Agency (IEA) made public that, during the year of 2010, despite the global economic crisis, a greenhouse gas emissions record was reached [35]. This means that the consequences of climate change are increasingly inevitable, and all of the possible strategies to reduce these emission levels should be established urgently. This chapter presents a methodology for a low-cost method that is applicable to reduce losses in parallel distribution transformer systems and studies the savings potential in several real situations [36].

Distribution transformer losses can be classified into no load losses and load losses.

No Load Losses

When a transformer is connected to a voltage level, losses occur. These losses are not dependent on the transformer power demand; hence, they are called no load losses.

The value of no load losses is constant, consisting of five distinct components, which are shown in Table 1.

Together, the hysteresis losses and eddy current losses account for 99% of the total losses, so that they are often the only losses taken into account, while the others are considered to be approximately equal to zero.

Table 1. Power transformers losses

Component	Percentage
Hysteresis losses in the core lamination and Eddy current losses in	=99%
the core lamination	<0.50% ≈ 0
Dielectric losses	<0.35% ≈ 0
Stray eddy current losses and core clamps, bolts and components	<0.15% ≈ 0
I^2R losses associated with no load current	

Load Losses

When the power demand on the transformer is varied, several losses result, including that due to component heat losses and that caused by the load current and eddy currents. These losses vary with temperature in addition to the conductivity characteristics of the materials. The most important value is due to losses characterized by I^2R.

Losses in Parallel Transformer Systems

Facilities are typically constructed of two transformers in parallel to ensure continuity in supply, in anticipation of demand expansion, to facilitate maintenance and due to regulations. In the parallel case, three operation modes can occur:

1. The demand power is less than the rated power of the first transformer.
2. The demand power is less than the rated power of the second transformer.
3. The demand power is higher than the nominal capacity of either of the two transformers.

These load situations present regularly within a given utility according to the demand curve of the installation, although the transformers are connected the same regardless of the load. As stated before, many different facility typologies, including industrial, commercial and health facilities, use more than one transformer to ensure continuous operation.

Taking into account the types of losses set out above, in the case of a system consisting of a transformer with a nominal power S_N, characterized by their losses in vacuum (L_0) with a percentage short circuit impedance (U_{SC}), the total losses when subjected to an apparent power demand S_L can be calculated as:

$$L_T = L_0 + \left(\frac{S_L}{S_N}\right)^2 U_{SC} \tag{1}$$

In the case of two transformers connected in parallel with a percentage short-circuit impedance test characteristics (U_{SC}), the load is distributed between both transformers, so that the lower is the value of U_{SC} and the short circuit currents are higher.

The total load power S_L is distributed between transformers 1 and 2 according to Eqs. (2) and (3):

$$S_{L,1} = \frac{S_L \cdot S_{N,1} \cdot U_{SC,T}}{\left(S_{N,1} + S_{N,2}\right) \cdot U_{SC,1}} \tag{2}$$

$$S_{L,2} = \frac{S_L \cdot S_{N,2} \cdot U_{SC,T}}{\left(S_{N,1} + S_{N,2}\right) \cdot U_{SC,2}} \tag{3}$$

$U_{SC,T}$ represents the short circuit impedance, which is determined according to Eq. (4):

$$U_{SC,T} = \frac{S_{N,1} + S_{N,2}}{\frac{S_{N,1}}{U_{SC,1}} + \frac{S_{N,2}}{U_{SC,2}}} \tag{4}$$

Hence, each transformer can be characterized by its percentage distribution of load PL1 and PL2 as shown in Eqs. (5) and (6):

$$PL_1 = \frac{S_{L,1}}{S_{N,1}} \tag{5}$$

$$PL_2 = \frac{S_{L,2}}{S_{N,2}} \tag{6}$$

Therefore, losses for a system operating in parallel are calculated according to Eq. (7):

$$L_{T,par} = L_{0,1} + \left(\frac{S_{L,1}}{S_{N,1}}\right)^2 U_{SC,1} + L_{0,2} + \left(\frac{S_{L,2}}{S_{N,2}}\right)^2 U_{SC,2} \tag{7}$$

METHODOLOGY AND STUDY

In this study, a study of losses in facilities with parallel transformers is carried out. Combinations of parallel transformers with powers from 100 kVA to 1,600 kVA were studied so that transformation facilities ranging from 200 kVA to 3,200 kVA could be investigated. The rated power ranks that were studied here cover the commercial, tertiary and industrial systems. Each transformer power was evaluated for transformers of three classes of efficiency, determined according to the norm EN 50464-1 as high, medium and low losses [37].

For each power of transformation and each load level, the load losses were computed and the parameter Parallel Losses Optimization (PLO) was calculated. This parameter represents the point of rupture that minimizes the losses in parallel transformers, opting for connection of either one or both of the two transformers.

Twelve points of consumption were tracked with a 434 Fluke Three Phase Power Quality Analyzer [38] over a month to obtain the power demand throughout the day. Using these data, demand curves were calculated. The reductions in the transformer losses were calculated by applying the method proposed in the chapter, and the energy savings in a given period was calculated to validate the PLO method.

A survey of several companies of maintenance of electric power systems over 1,000 transformer systems with rated power values equal to those used in the studied the chapter was carried out to determine the maintenance protocols. This background investigation demonstrated that maintenance is not usually rigorous and losses in transformers are considered an inevitable component of the distribution system.

The Transformer Facilities Studied

To carry out the study, transformers with the various characteristics and rated power values were combined, based on the values given in Table 2. These combinations led to 13 rated powers with 3 levels of energy efficiency, so that a total of 39 study cases were considered (Table 3). The analyzed rated powers correspond to typical cases used in a wide range of applications. All transformer data corresponds to real equipment, and the manufacturer specified loss data were used.

Table 2. Characteristics of the power transformers studied

S_N (kVA)	EN50464-1 classification	U_{SC} (%)	P_0 (W)
100	E_0D_k - High	4	320
160	E_0D_k - High	4	460
250	E_0D_k - High	4	650
400	E_0D_k - High	4	930
630	E_0D_k - High	4	1,200
1,000	E_0D_k - High	5	1,700
1,600	E_0D_k - High	6	2,600
100	D_0C_k - Medium	4	260
160	D_0C_k - Medium	4	375
250	D_0C_k - Medium	4	530
400	C_0B_k - Medium	4	750
630	D_0C_k - Medium	4	940
1,000	D_0C_k - Medium	5	1,400
1,600	D_0C_k - Medium	6	2,200
100	C_0B_k - Low	4	210
160	C_0B_k - Low	4	300
250	C_0B_k - Low	4	425
400	C_0B_k - Low	4	610
630	C_0B_k - Low	4	800
1,000	C_0B_k - Low	5	1,100
1,600	C_0B_k - Low	6	1,700

Table 3. The power facilities studied

S_N (kVA)	Parallel transformers	EN50464-1 classification
200	2 x 100	E_0D_k, D_0C_k, C_0B_k
260	100 +160	E_0D_k, D_0C_k, C_0B_k
320	2 x 160	E_0D_k, D_0C_k, C_0B_k
410	160 + 250	E_0D_k, D_0C_k, C_0B_k
500	2 x 250	E_0D_k, D_0C_k, C_0B_k
650	250 + 400	E_0D_k, D_0C_k, C_0B_k
800	2 x 400	E_0D_k, D_0C_k, C_0B_k
1,030	400 + 630	E_0D_k, D_0C_k, C_0B_k
1,260	2 x 630	E_0D_k, D_0C_k, C_0B_k
1,630	630 + 1,000	E_0D_k, D_0C_k, C_0B_k
2,000	2 x 1,000	E_0D_k, D_0C_k, C_0B_k
2,600	1000 + 1,600	E_0D_k, D_0C_k, C_0B_k
3,200	2 x 1,600	E_0D_k, D_0C_k, C_0B_k

Calculation of the Losses and Determination of the PLO

The respective losses of each transformer and that of the transformers in parallel were plotted to yield each point on the demand load curve, based on Eq. (8):

$$S_N = S_{N,1} + S_{N,2} \tag{8}$$

Depending on the load and using the method given in Eqs. (1) and (7), the losses were calculated and the PLO were determined. The PLO power is based on the system loss determinations and can recommend connection of both transformers even when the rated power in one transformer has not been reached ($S_{N,1}$). This strategy allows the losses to be minimized based on the minimum losses of operation instead of the rated powers.

Analysis of the Maintenance Protocols

The consulted transformation facilities use the powers outlined in Table 3, classified according to the type of energy efficiency level (Table 4). The maintenance operations were classified into four basic types:

- Only oil maintenance: consists of the verification of the oil levels and physico-chemical parameters.
- Yearly verification: annual verification of the absence of defects in transformers, including visual inspection and analysis of the oil.
- Earth fault verification: control of the absence of dangerous leakage currents in the transformation utility.
- Advanced maintenance: advanced maintenance, which includes control of losses, verification of liquid and isolation levels.

Table 4. The maintenance survey details

S_N (kVA)	Number	E_0D_k	D_0C_k	C_0B_k
200	84	63	17	4
260	88	83	4	1
320	85	65	17	3
410	86	6	4	6
500	87	81	6	0
650	93	80	9	4
800	91	69	18	4
1,030	65	56	8	1
1,260	89	78	8	3
1,630	97	85	9	3
2,000	68	51	11	6
2,600	67	56	2	9
Total	1,000	843	113	44

RESULTS AND ANALYSIS

Using the methodology described in this chapter, the points of loss optimization were calculated in addition to monitoring of the installations and estimation of the potential savings. Finally, the results of the compiled information regarding the transformer utilities are presented for analysis of the potential to use the proposed methodology.

The Optimization Point Calculation

The transformers systems for each of the powers and efficiency classes presented in the previous section were investigated. To combine the transformers within the utility, the losses were simulated at different load levels (S_L), working with each of the transformers and simultaneous operation obtaining a load versus loss graph as shown in Figure 1. This figure illustrates the result for a transformation of 410 kVA power (transformers of 160 + 250 kVA, efficiency E_0D_k class).

In Figure 1, the losses of the first transformer operating alone can be seen in addition to the losses of the second transformer under the same strategy of operation, the loss of the system in parallel and the curve of minimal losses (optimal). This figure can be used to determine the point PLO as the power load value at which the losses decrease for parallel connection of both transformers (the details can be seen in Figure 2).

The optimal work with the connection of a single transformer system can be seen in area A, while the optimal work with a parallel connection is shown in area B. The PLO was calculated for each of the cases, yielding the values indicated in Table 5. It should be noted that, as a general rule, as the efficiency of the transformers increases, the PLO decreases even though the values are very similar for all efficiency levels. The effect of the increase on the efficiency level is a reduction of the losses for any operating condition.

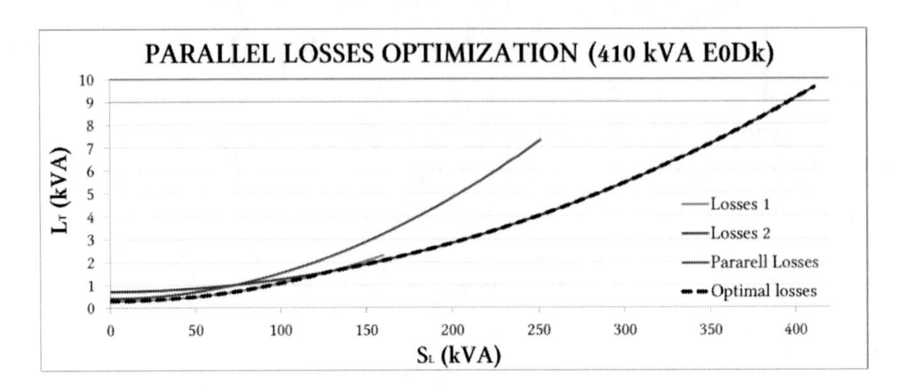

Figure 1. The losses versus transformer load.

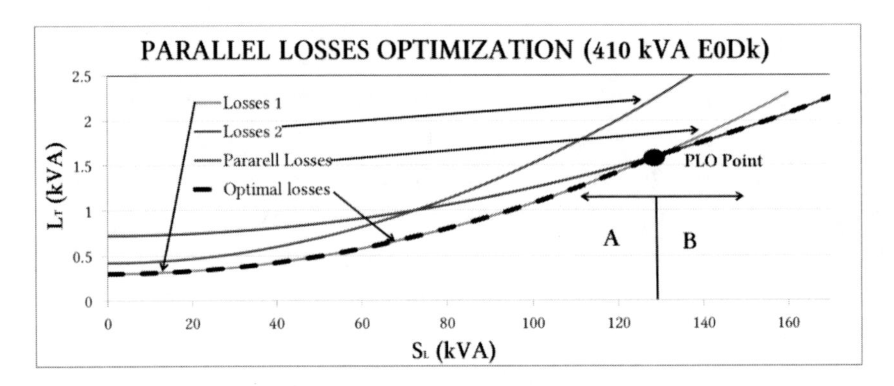

Figure 2. The PLO detail.

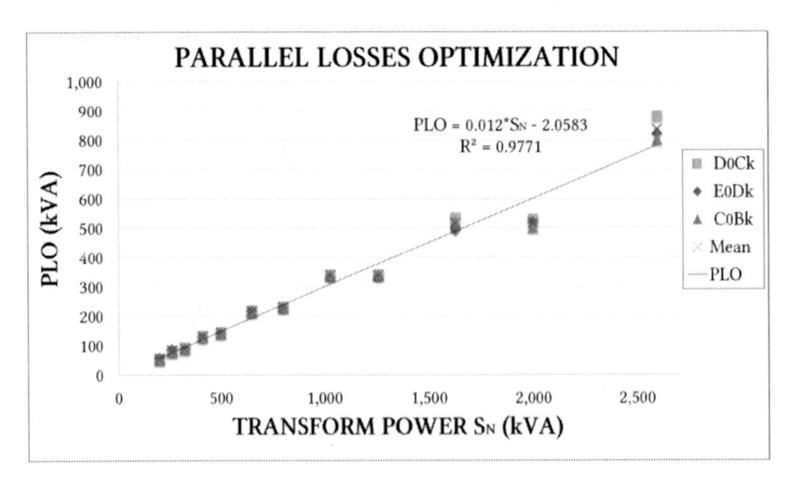

Figure 3. The PLO and S_N correlation.

Table 5. The PLO results

S_N (kVA)	E_0D_k	D_0C_k	C_0B_k	Mean
200	56	55	54	55
260	86	82	79	82
320	89	92.8	89	90
410	128	134	129	130
500	142.5	145	140	143
650	217	220	212	216
800	228	232	229	230
1,030	332	340	340	337
1,260	333	340	340	338
1,630	491	535	522	516
2,000	520	530	500	517
2,600	820	880	800	833

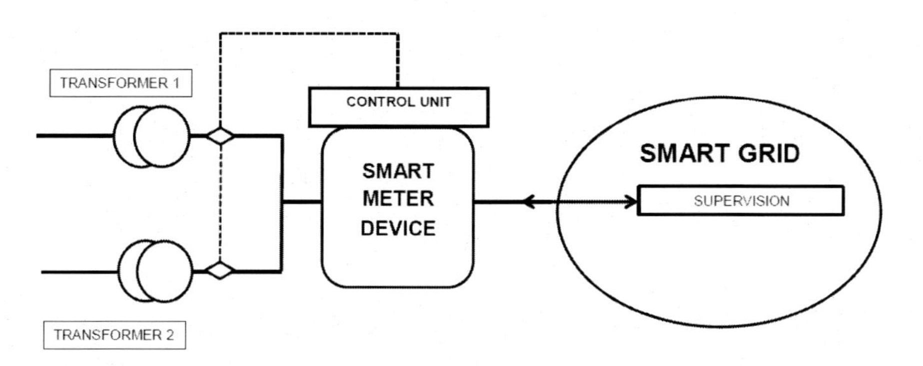

Figure 4. The PLO system in smart grids.

A graphic representation of the values is given in Figure 3, including the results of a statistical treatment of these numbers that includes a correlation, which allows calculation of the PLO point for a rated power S_N.

For each efficiency level shown, both the PLO value and the average value of the PLO for each S_N rated power are given. Based on the results of this work, a new method for the optimization of parallel transformers is given. Conducting a further analysis to validate the statistical correlation allows for a proper linear correlation to be extracted from the probed values with a level of adjustment $R^2 = 0.9771$. The calculation method used is based simply on the rated power, as given in Eq. (9):

$$PLO = 0.012 \cdot S_N - 2.0583 \quad (9)$$

Through application of the proposed Eq., the proposed method enables implementation of the system in any existing or future installation, regardless of the system's characteristics. The simplicity of the algorithm allows for programming with a programmable automaton, which will control the attach/detach commands for the parallel transformers. Due to the need to adapt systems of electrical measurement to the requirements for new Smart Grid systems, the measurement devices used in power distribution systems within US and Europe must be completely changed with a maximum time horizon of 2020. Smart Metering systems have outputs with custom programming capacity and in some cases can perform logic calculations. With a system like that shown in Figure 4, it is possible to use the PLO method proposed in this chapter with a cost of almost zero to provide returns on the investment with in a very low period. The proposed strategy also can be used in a more significant way in future power transformer systems for reducing losses and can be integrated as a common algorithm in transformer systems.

The capacity for remote management of the Smart Meter allows, through the use of Power Line Communication (PLC) or any other remote system monitoring, controlling and updating of the installation.

Energy Savings in the Monitored Facilities. Experimental Validation

Power Quality Analyzers were placed for a month in twelve facilities of commercial, educational, sports and industrial type with respective powers of 650 kVA, 1,260 kVA, 1,630 kVA and 2,600 kVA to obtain their typical load profiles during the day. The simulated configurations are detailed in Table 6. Three types of efficiency levels were studied for each power installation. The sampling was made at a frequency of 1 Hz to prevent exhaustion of the characteristics of the device's memory. This sampling rate is generally too low for monitoring power quality but was sufficient for the purpose of the study. All facilities had switching systems that would perform a switching operation between one and another transformer or both in parallel. The monitored demand curves show the apparent power consumed at each moment (S_{dem}). Next, for each instant of time, losses in the installation with both processors in parallel ($L_{T,par}$) and optimal configuration ($L_{T,opt}$) and the PLO for each case were computed.

The results are presented in Figure 5a and 5b in addition to Figure 8a and 8b with details of losses in the E_0D_k efficiency class utilities.

Based on the monthly demand graph and the annual operational data, the energy losses were calculated for a year, both in the mode of operation and in the parallel optimized mode. Figures 9-12 shows the kWh losses throughout the year with the current operation, the losses with the optimized mode and the savings percentage. These savings percentage values are summarized in Table 7. The reductions in annual system transformation losses range, depending on the load curve, between 5.97% and 41.46%. These reductions in losses represent a significant percentage savings without the need to modify or replace the transformers and their associated systems.

Table 6. The experimentally studied power transformers

S_N (kVA)	Parallel transformers	EN50464-1 classification
650	250 + 400	E_0D_k, D_0C_k, C_0B_k
1,260	2 x 630	E_0D_k, D_0C_k, C_0B_k
1,630	630 + 1,000	E_0D_k, D_0C_k, C_0B_k
2,600	1,000 + 1,600	E_0D_k, D_0C_k, C_0B_k

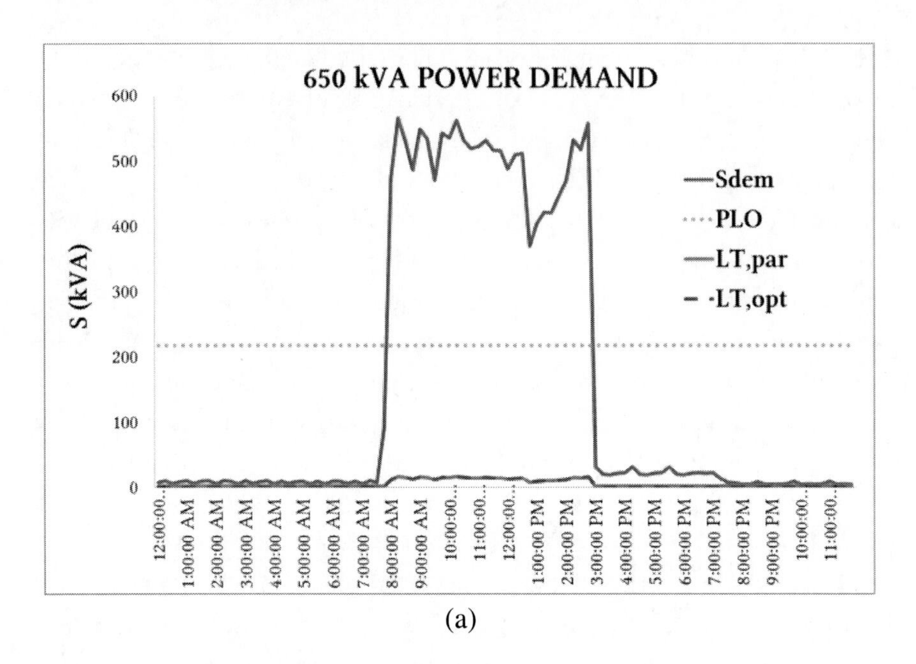

(a)

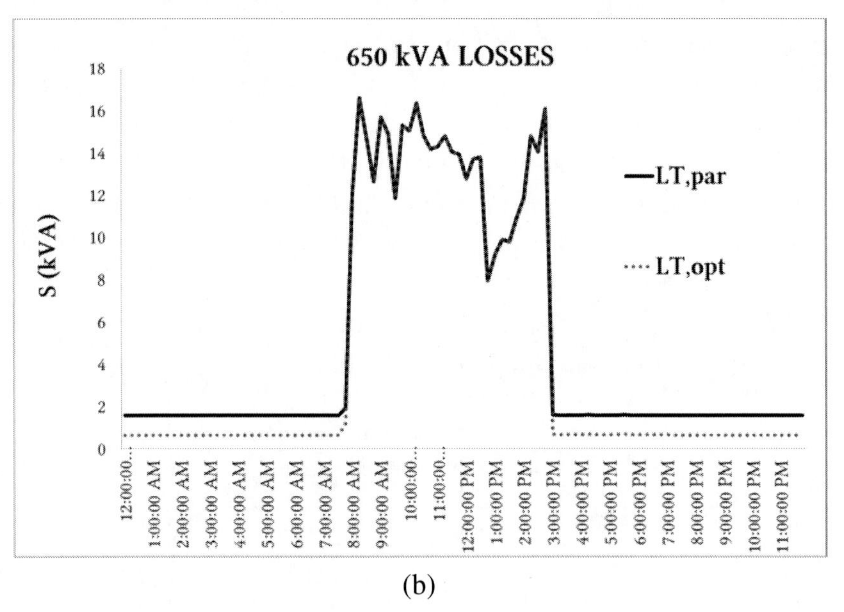

(b)

Figure 5a and 5b. The 650 kVA case study demand curve and transformer losses.

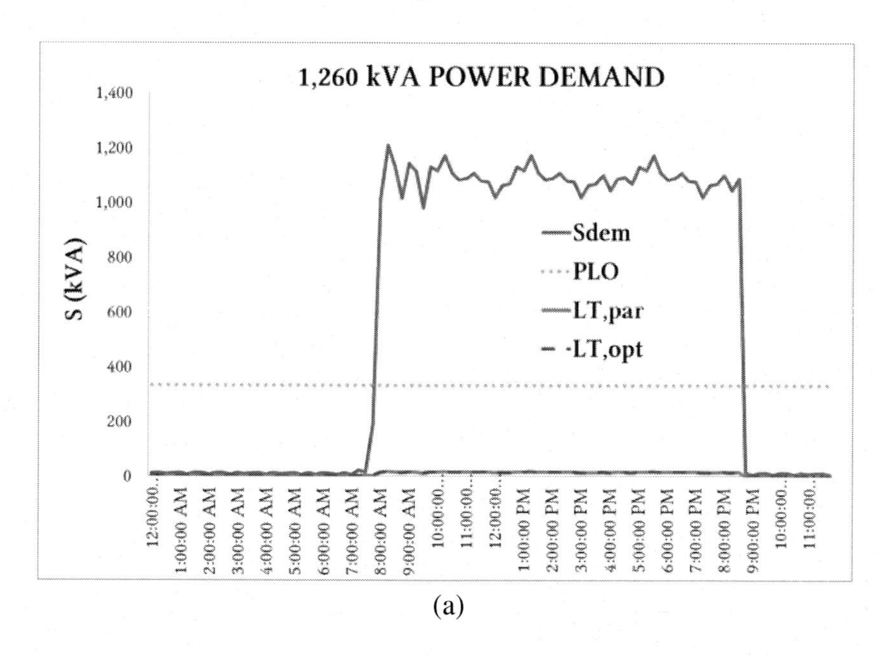

(a)

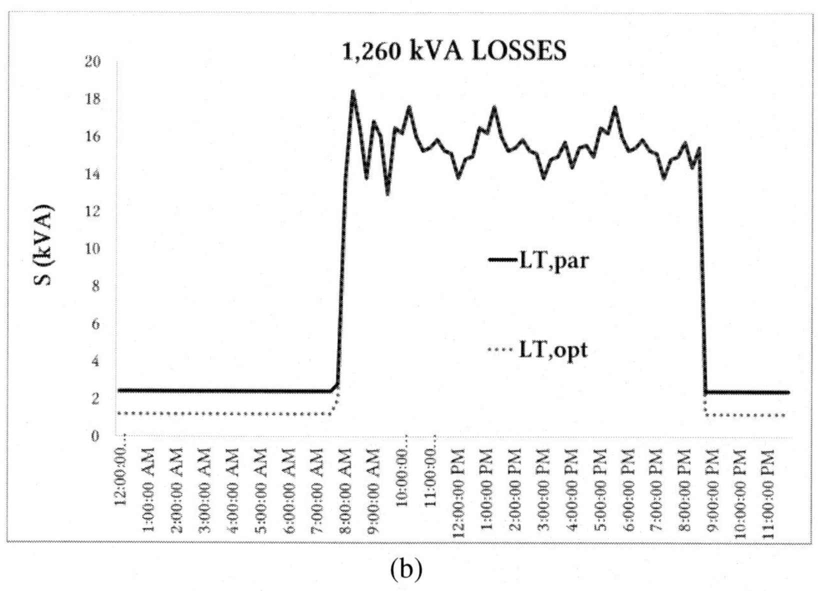

(b)

Figure 6a and 6b. The 1,260 kVA case study demand curve and transformer losses.

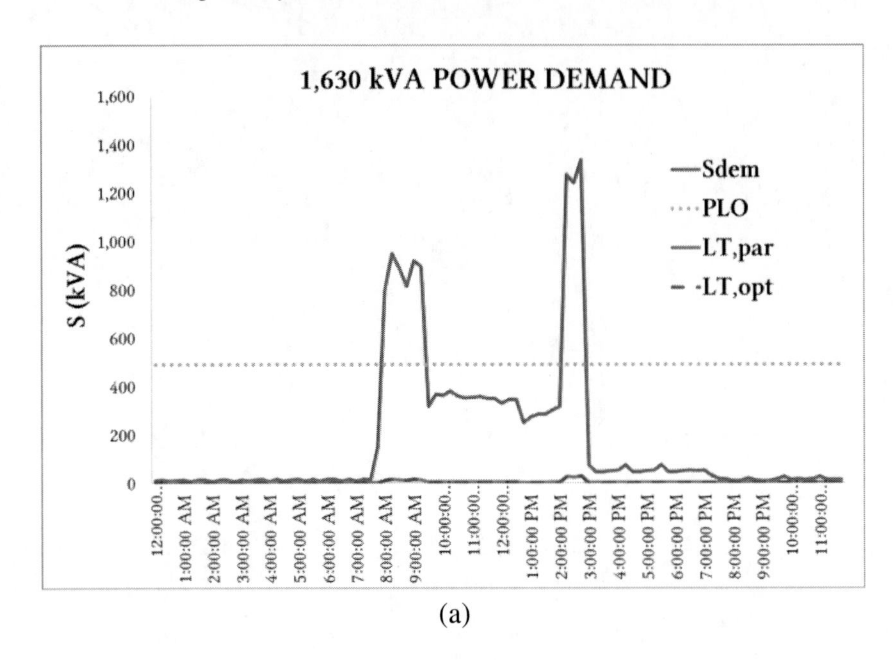

(a)

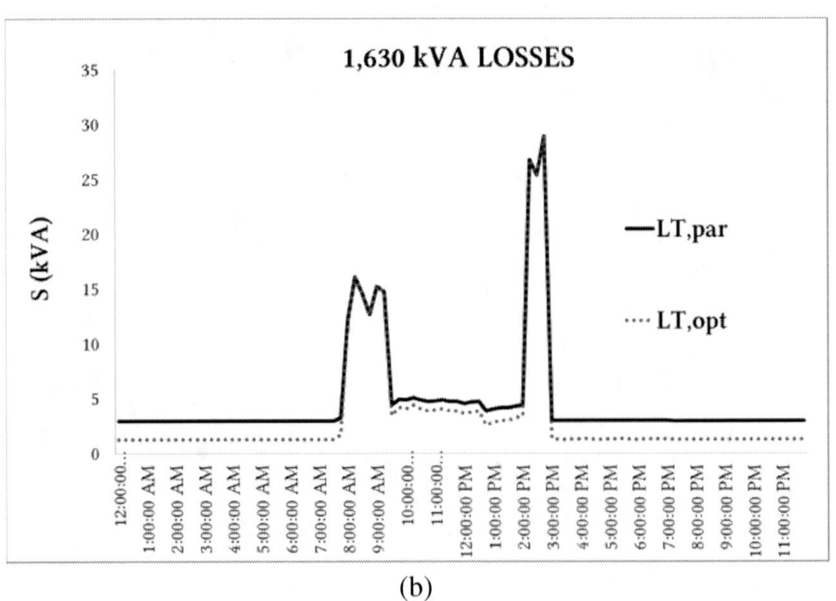

(b)

Figure 7a and 7b. The 1,630 kVA case study demand curve and transformer losses.

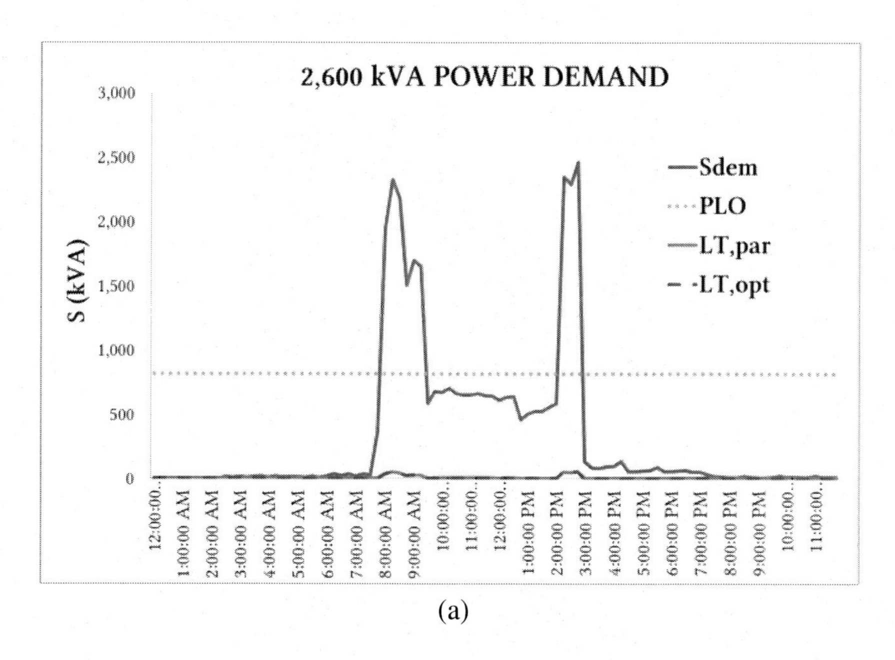

(a)

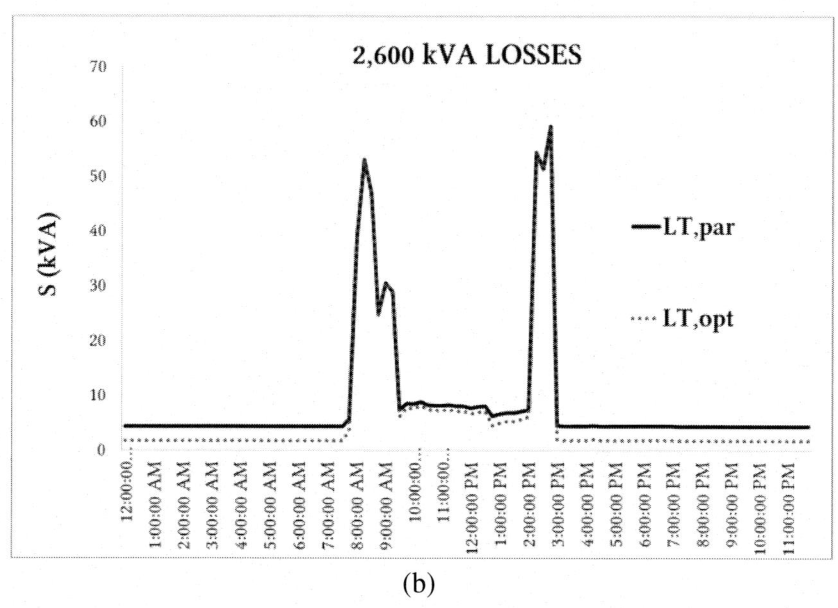

(b)

Figure 8a and 8b. The 2,600 kVA case study demand curve and transformer losses.

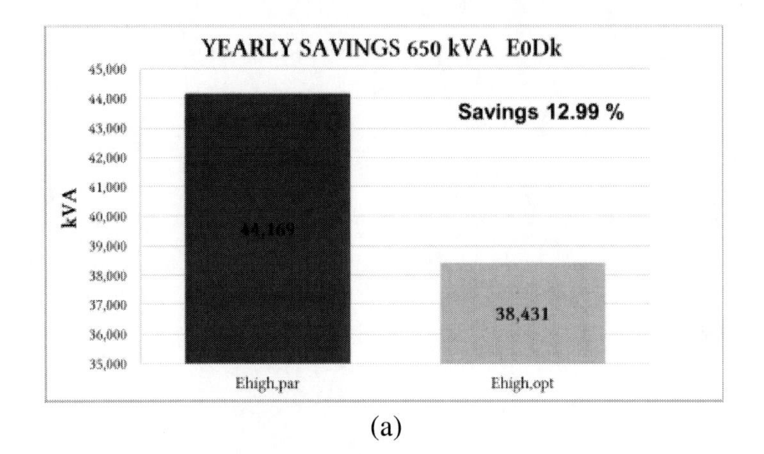

(a)

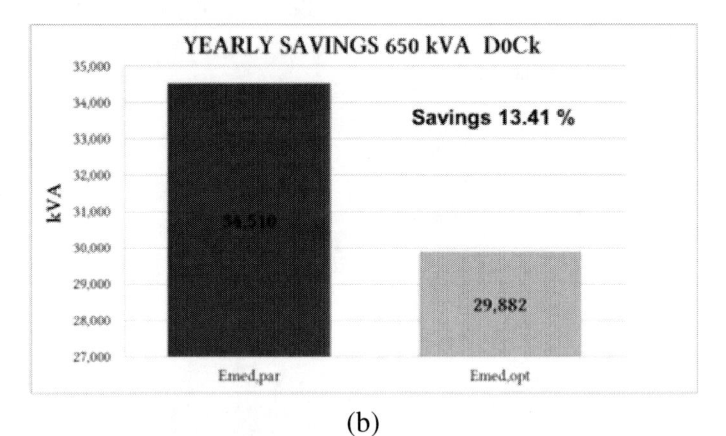

(b)

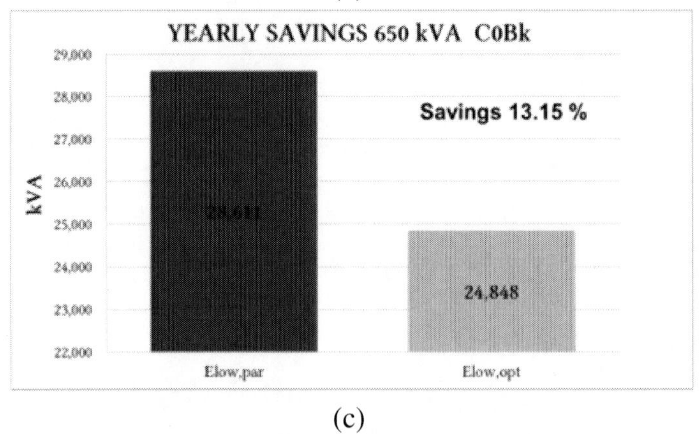

(c)

Figure 9. The 650 kVA case study transformer losses reduction.

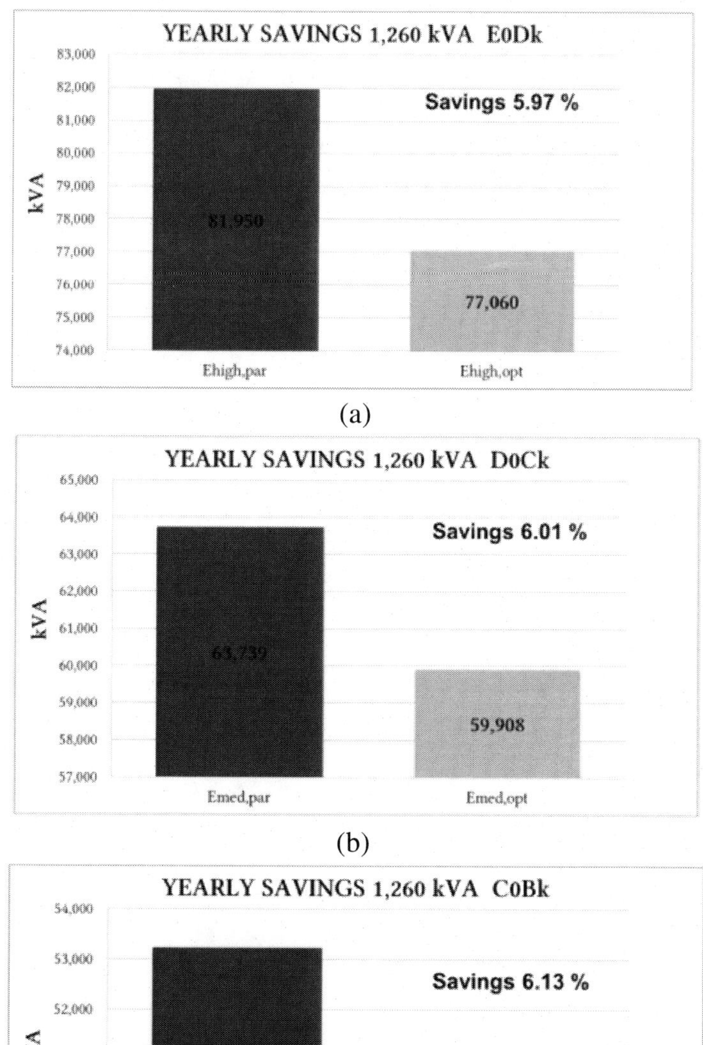

Figure 10. The 1,260 kVA case study transformer losses reduction.

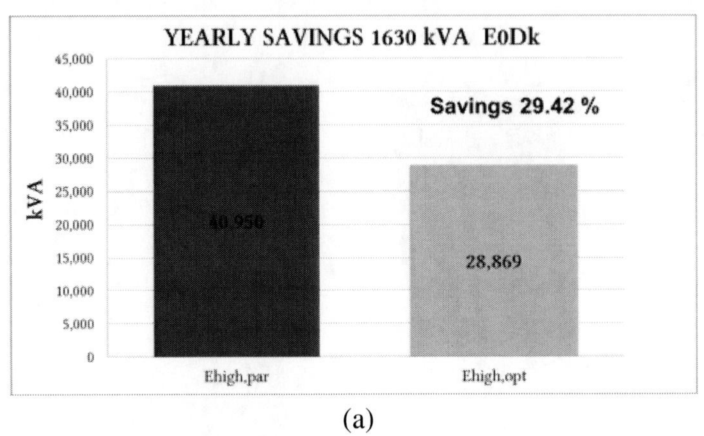

(a)

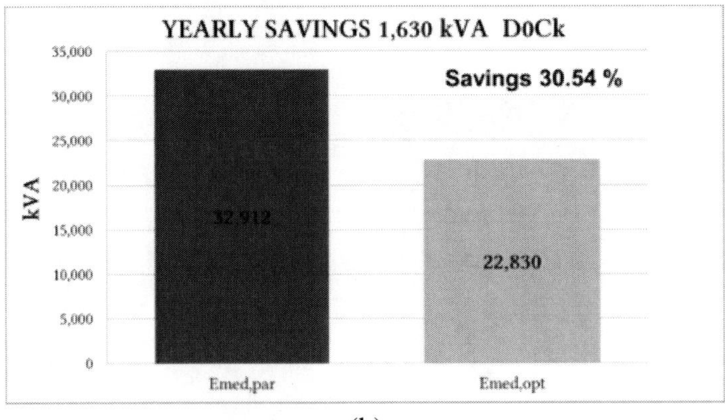

(b)

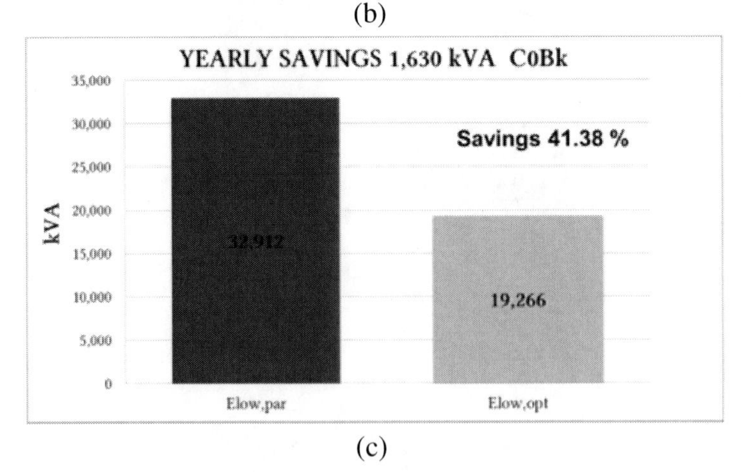

(c)

Figure 11. The 1,630 kVA case study transformer losses reduction.

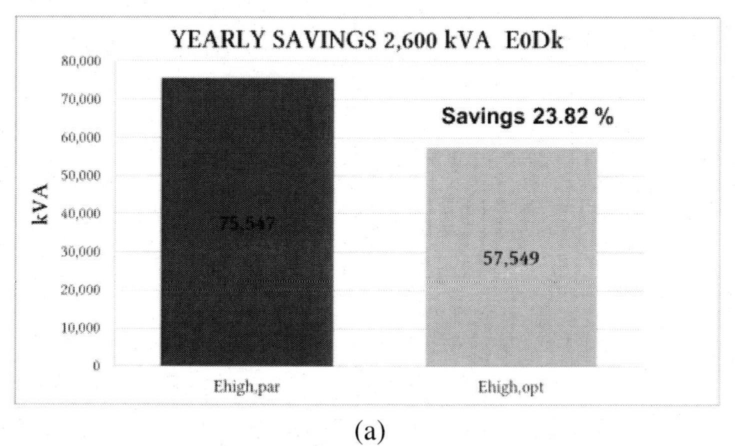

(a)

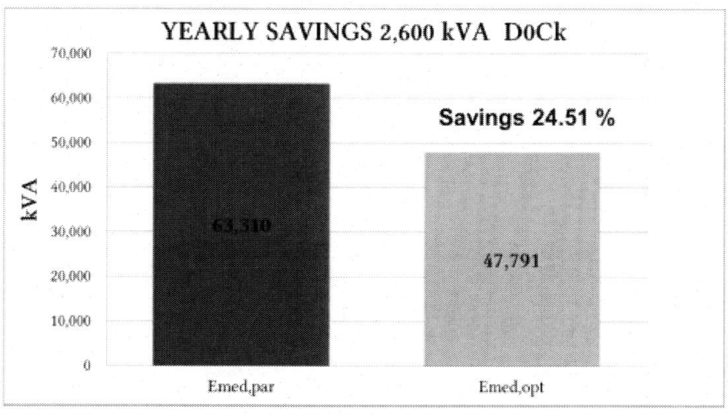

(b)

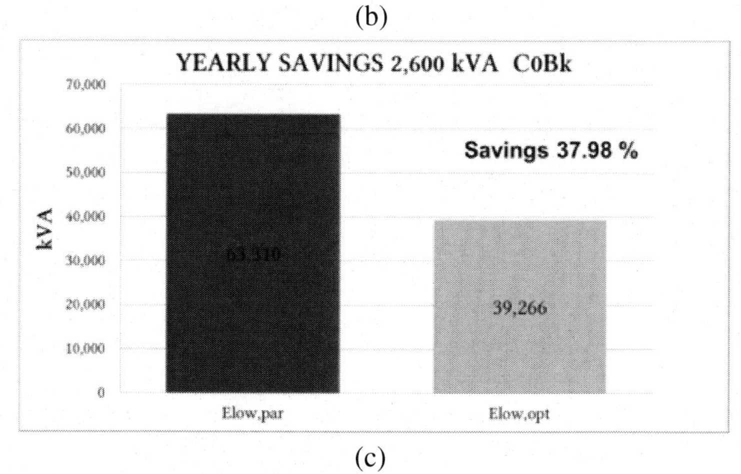

(c)

Figure 12. The 1,630 kVA case study transformer losses reduction.

Table 7. The yearly percentage of energy savings from losses

S_N (kVA)	Parallel transformers	% Savings (E_0D_k)	% Savings (D_0C_k)	% Savings (C_0B_k)
650	250 + 400	12.99	13.41	13.15
1,260	2 x 630	5.97	6.01	6.13
1,630	630 + 1,000	29.50	30.63	41.46
2,600	1,000 + 1,600	23.82	24.51	37.98

Experimental Data of the Maintenance Operations

Electrical systems operation and maintenance companies provided data from transformation utilities, efficiency levels and annual maintenance protocols. The details of the facilities and their efficiency levels are given in Table 8 and Figure 13.

Table 8. The details of the studied transformer installations

S_N (kVA)	NUMBER	E_0D_k	D_0C_k	C_0B_k
200	84	63	17	4
260	88	83	4	1
320	85	65	17	3
410	86	76	4	6
500	87	81	6	0
650	93	80	9	4
800	91	69	18	4
1,030	65	56	8	1
1,260	89	78	8	3
1,630	97	85	9	3
2,000	68	51	11	6
2,600	67	56	2	9
TOTAL	1,000	843	113	44
%		84%	11%	4%

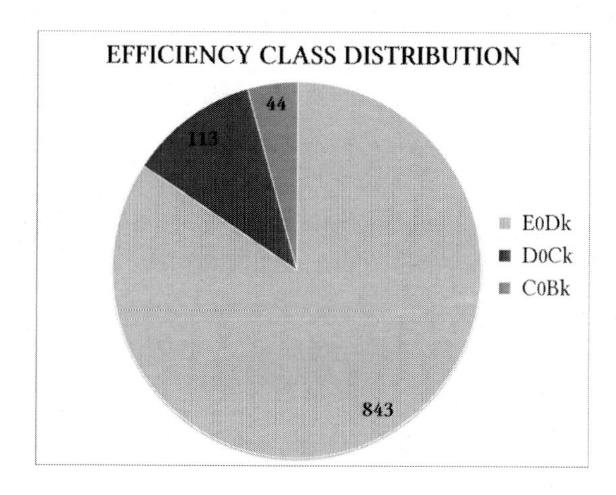

Figure 13. The efficiency class of the case studies.

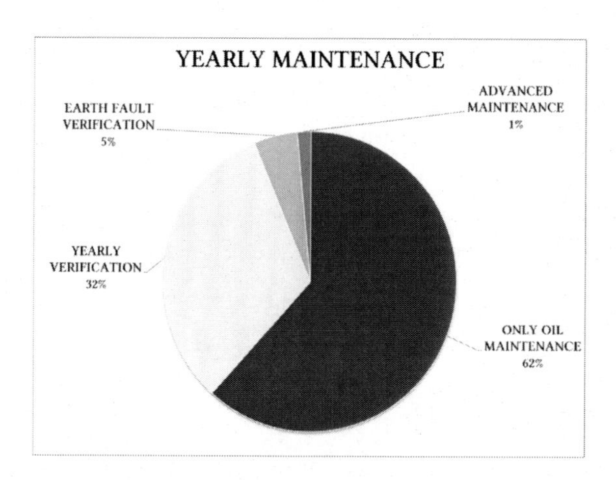

Figure 14. The maintenance protocols used in the studied cases.

In other electric machines, such as drivers or generators, companies use advanced maintenance and failure prevention models [39, 40]. The data show that, despite the importance of transformers as elements that generate high losses in distribution systems, they are not usually installed using methods that evaluate energy losses throughout the transformers' life cycles. Methods such as TOC evaluate these parameters. Only 5% of facilities demonstrated the highest level of energy efficiency. Additionally, the survey denoted that approximately 1% was subjected to advanced maintenance oriented to the reduction of losses (Figure 14). This figure shows that transformer utilities are not usually considered properly in energy optimization strategies, even though there is a high potential for GHG emissions reduction in addition to the associated economic losses. Both electrical companies and the final energy consumers can improve the losses in the global system by reducing the losses associated with their power transformers. The operation and maintenance of transformers also should be established as a key point due to the high energy losses associated with these systems.

CONCLUSION

This chapter analyzes the energy losses in transformation systems composed of transformers in parallel and proposes a method, called PLO, that allows reductions in these losses. Distribution transformer system losses represent an important contribution to the quantity of GHG emitted to the atmosphere and have a high economic cost. In the chapter, a method is proposed and validated for transformer utilities of three efficiency levels. This method can be implemented with any transformer, independent of its characteristics. New low-loss transformers have lower losses but a higher economic cost, so that their installation is cost-effective only if is the total cost over the life cycle is lower. However, replacement of existing transformers is rarely profitable. The proposed method reduces losses throughout the life cycle in new or existing installations. For implementation, the system proposed in this chapter does not require any additional devices and allows energy savings of up to 41% to be obtained over the initial losses when it is used in parallel transformers; the study also proposes use of an automated system instead of manual disconnection. The new energy measurement equipment in smart grid systems facilitates the installation and operation of this method. In facilities with no continuous hourly operation, the savings percentage may be significantly higher because at night or on weekends the transformers are connected in parallel, thereby initiating losses. The connection and disconnection of parallel transformers, however, according to the PLO, reduces these losses. In facilities with seasonal operation, the annual losses can be reduced very significantly by performing the optimization protocol described in this study.

The load-demand curves of 12 facilities with transformers with different levels of efficiency were monitored to analyze their respective load profiles and to detect when energy savings can be achieved by means of connection modification. Applying the proposed PLO method, loss reductions were achieved in all cases studied. Data from one thousand actual installations demonstrate the need to implement systems and protocols for loss reduction, as proposed here. The potential to reduce cost and GHG emissions has motivated UNED to obtain a patent pending status for the system, with number P201101267.

REFERENCES

[1] Kennedy B. W. *Energy efficient transformers*. New York: McGraw-Hill; 1998.

[2] Hasegawa R., Azuma D. Impacts of amorphous metal-based transformers on energy efficiency and environment. *J Magn Magn Mater* 2008;320:2451–6.

[3] Acharya N., Mahat P., Mithulananthan N. An analytical approach for DG allocation in primary distribution network. *Int J Electr Power Energy Syst* 2006;28:669–678.

[4] Olivares J. C., Yilu L., Canedo J. M., Escarela-Perez R., Driesen J., Moreno P. Reducing losses in distribution transformers. *IEEE Trans Power Deliv* 2003;18:821–6.

[5] Hasegawa R., Azuma D. Impacts of amorphous metal-based transformers on energy efficiency and environment. *J Magn Magn Mater* 2008;320:2451–6.

[6] Georgilakis P. S. Differential evolution solution to transformer no-load loss reduction problem. *IET Gener, Trans Distrib* 2009;3:960–9.

[7] Olivares-Galván J. C., Georgilakis P. S., Ocon-Valdez R. A review of transformer losses. *Electr Power Compon Syst* 2009;37:1046–62.

[8] Chiradeja P, Ramakumar R. An approach to quantify the technical benefits of distributed generation. *IEEE Trans. Energy Conversion,* 2004;19(4):764- 773.

[9] Nasser G. A., Kurrat M. Efficient integration of distributed generation for meeting the increased load demand. *Int J Electr Power Energy* Syst, 2011; 33(9):1572-1583.

[10] Nochumson C. J. Considerations in application and selection of unit substation transformers. *IEEE Trans Ind Appl* 2002;38:778–87.

[11] Rasmusson P. R. Transformer economic evaluation. *IEEE Trans Ind Appl* 1984;20:355–63.

[12] Bins D. F., Crompton A. B., Jaberansari A. Economic design of a 50 kVA distribution transformer. Part 2: effect of different core steels and loss capitalisations. *IEEE Proc Gener, Trans, Distrib* 1986;133: 451–6.

[13] ANSI/IEEE. *IEEE loss evaluation guide for power transformers and reactors*. ANSI/IEEE Standard C57.120-1991; 1992.

[14] Nickel D. L., Braunstein H. R. Distribution transformer loss evaluation: I –proposed techniques. *IEEE Trans Power Appar Syst* 1981;100:788–97.

[15] Nickel D. L., Braunstein H. R. Distribution transformer loss evaluation: II – load characteristics and system cost parameters. *IEEE Trans Power Appar Syst* 1981;100:798–811.

[16] Merritt S., Chaitkin S. No load versus load loss. *IEEE Ind Appl Mag* 2003;9:21–8.

[17] Georgilakis P. S. Decision support system for evaluating transformer investments in the industrial sector. *J Mater Process Technol* 2007;181:307–12.

[18] Baranowski J. F., Hopkinson P. J. An alternative evaluation of distribution transformers to achieve the lowest TOC. *IEEE Trans Power Deliv* 1992;7:614–9.

[19] Kang M., Enjeti P. N., Pitel I. J. Analysis and design of electronic transformers for electric power distribution system, *IEEE Trans. Power Electr.* 1999;14:1133–1141.

[20] Ghosh S., Ghoshal S. P., Ghosh S. Optimal sizing and placement of distributed generation in a network system. *Int J Electr Power Energy Syst,* 2010; 32:849–856.

[21] Moses M., Masoum M., Toliyat H. Dynamic modeling of three-phase asymmetric power transformers with magnetic hysteresis: no-load and inrush conditions. *IEEE Trans. Energy Conversion,* 2010;25(4): 1040–1047.

[22] McMurray W. Power converter circuits having a high frequency link, *US Patent 3*,517,300, June 23, 1970.

[23] Venkataramanan G., Johnson B. K., Sundaram A. An ac–ac power converter for custom power applications. *IEEE Trans. Power Delivery* 1996; 11:1666–1671.

[24] Brooks J. L. Solid State Transformer Concept Development, Naval Material Command, Civil Engineering Laboratory, *Naval Construction Battalion Ctr.*, Port Hueneme, CA, 1980.

[25] EPRI Report, Proof of the principle of the solid-state transformer: the AC/AC switch mode regulator, *EPRI TR-105067, Research Project* 8001-13, Final Report, August 1995.

[26] Harada K., Anan F., Yamasaki K., Jinno M., Kawata Y., Nakashima T., Murata K., Sakamoto H. Intelligent transformer, in: *Proceedings of the 1996 IEEE PESC Conference 1996:*1337–1341.

[27] Manjrekar M. D., Kieferndorf R., Venkataramanan G. Power electronic transformers for utility applications, in: *Proceedings of the 2000 IEEE IAS Annual Meeting* 2000:2496–2502.

[28] Ronan E. R., Sudhoff S. D., Glover S. F., Galloway D. L. A power electronic based distribution transformer, *IEEE Trans. Power Delivery* 2002;17:537–543.

[29] Marchesoni M., Novaro R., Savio S. AC locomotive conversion systems without heavy transformers: is it a practicable solution, in: *Proceedings of the 2002 IEEE International Symposium on Industrial Electronics* 2002:1172–1177.

[30] Targosz R. The potential for global energy savings from high energy efficiency distribution transformers. *Leonardo Energy,* 2005. <http://www.leonardoenergy.org/repository/Library/Reports/Transformers-Global.pdf> [accessed 05.11].

[31] Zhang Y-J., Wei Y-M. An overview of current research on EU ETS: evidence from its operating mechanism and economic effect. *Appl Energy* 2010;87:1804–14.

[32] Chen W. T., Li Y. P., Huang G. H., Chen X., Li Y. F. A two-stage inexact-stochastic programming model for planning carbon dioxide emission trading under uncertainty. *Appl Energy* 2010;87:1033–47.

[33] Vad Mathiesen B., Lund H., Karlsson K. 100% renewable energy systems, climate mitigation and economic growth. *Appl Energy* 2011;88:488–501.

[34] Irrek W., Topalis F., Targosz R., Rialhe A., Frau J. Policies and measures fostering energy-efficient distribution transformers. *Report of European Commission Project No EIE/05/056/SI2.419632*; June 2008.

[35] CO_2 emissions reach a record high in 2010; 80% of projected 2020 emissions from the power sector are already locked in. *International Energy Agency*, 2011. <http://www.iea.org> [accessed 05.11].

[36] Méndez V. H., Rivier J., de la Fuente J. I., Gomez T., Arceluz J., Marín J. Impact of distributed generation on distribution investment deferral. *Int J Electrical Power Energy Syst* 2006;28(4):244–252.

[37] CENELEC. Three-phase oil-immersed distribution transformers 50 Hz, from 50 kVA to 2500 kVA with highest voltage for equipment not exceeding 36 kV –Part 1: General requirements. *CENELEC Standard EN* 50464-1; 2007.

[38] FLUKE 430 *Series Three-Phase Power Quality Analyzers*, 2011. <http://www.fluke.com/fluke/usen/power-quality-tools/three-phase/fluke-430-series.html> [accessed 05.11].

[39] Chuang F., Luqing Y., Yongqian L., Ren Y., Benoit L., Yuanchu C., Yuming Z. Predictive maintenance in intelligent-control-maintenance-management system for hydroelectric generating unit. *IEEE Trans. Energy Conversion*, 2004;19(1):179- 186.

[40] Alberti L., Fornasiero E., Bianchi N., Bolognani S. Rotor Losses Measurements in an Axial Flux Permanent Magnet Machine. *IEEE Trans. Energy Conversion*, 2011;26(2):639-645.

In: Renewable Electric Power Distribution … ISBN: 978-1-53614-202-0
Editors: A. Colmenar-Santos et al.

Chapter 4

ENERGY EFFICIENCY IMPROVEMENT IN POWER CONVERTERS

Ana-Rosa Linares-Mena[1,*], Jesús Fernández Velázquez[2] and Carlos de Palacio[3]

[1]Instituto Tecnológico y de Energías Renovables, S.A (ITER), S/C de Tenerife, Spain

[2]Agencia Insular de Energía de Tenerife (AIET), S/C de Tenerife, Spain

[3]ABB, Madrid, Spain

ABSTRACT

Grid connected energy storage systems are expected to play an essential role in the development of Smart Grids, providing, among other benefits, ancillary services to power grids. It is therefore crucial to design and develop control and conversion systems that represent the key instrument where intelligence for decision-making is applied, in order to validate and ensure its optimal operation as part and parcel of the

[*] Corresponding Author Email: ar.linares.mena@gmail.com.

electrical system. The present research describes the design and development of a battery energy storage system based on an AC-DC three-phase bidirectional converter capable of operating either in charge mode to store electrical energy, or in discharge mode to supply load demands. The design is modelled with MATLAB® Simulink® environment in order to evaluate the performance during load variations. Moreover, the assessment is complemented by a global sensitivity analysis for variations in the operating parameters set by the transmission system operator. The effectiveness of the simulation is confirmed by implementing the system and carrying out grid connection tests, obtaining efficiencies over 98% for values over the 30% of the bidirectional converter rated power.

Keywords: electrical energy storage, bidirectional converter, battery, hysteresis control

NOMENCLATURE

Acronyms

AC	Alternating Current
BESS	Battery Energy Storage Systems
CC	Current control
COP21	21st Conference of the Parties
DC	Direct Current
EPRI	Electric Power Research Institute
ES	Electric System
ESS	Energy Storage Systems
EU	European Union
HB	Hysteresis Band
IGBT	Insulated Gate Bipolar Transistor
ITER	Instituto Tecnológico y de Energías Renovables, S.A.
LHB	Lower Hysteresis Band
PEI	Power Electronic Interface
SPWM	Sinusoidal Pulse Width Modulation

RES	Renewable Energy Systems
RTU	Remote Terminal Unit
RMS	Root Mean Square Value
SCADA	Supervisory Control and Data Acquisition
SOC	State of Charge
TSO	Transmission System Operator
UHB	Upper Hysteresis Band
VSI	Voltage Source Inverter

Parameters

$i_{a,ref}$	Reference current in the phase a (A)
$i_{b,ref}$	Reference current in the phase b (A)
$i_{c,ref}$	Reference current in the phase c (A)
i_a	Measured current in the phase a (A)
i_b	Measured current in the phase b (A)
i_c	Measured current in the phase c (A)
v_a	Measured voltage in the phase a (V)
v_b	Measured voltage in the phase b (V)
v_c	Measured voltage in the phase c (V)
ε_a	Error signal in the phase a
ε_b	Error signal in the phase b
ε_c	Error signal in the phase c
S_a	Signal for commanding the transistor base drives in the phase a
S_b	Signal for commanding the transistor base drives in the phase b
S_c	Signal for commanding the transistor base drives in the phase c
$\bar{v}$	Voltage vector
$\bar{\iota}$	Current vector

$\bar{a}$	Bidimensional vector
I_{max}	Peak Current
θ	Angle between the voltage vector and real axis
$i_{a,up}$	Upper current band in the phase a (A)
$i_{a,low}$	Lower current band in the phase a (A)
$i_{b,up}$	Upper current band in the phase b (A)
$i_{b,low}$	Lower current band in the phase b (A)
$i_{c,up}$	Upper current band in the phase c (A)
$i_{c,low}$	Lower current band in the phase c (A)
h	Hysteresis Band limit
v_{DC}	Terminal voltage of the battery bank (V)
ω	Angular velocity (rad/s)
v_{ag}	Ground-to-line voltage in the phase a (V)
v_{bg}	Ground-to-line voltage in the phase b (V)
v_{cg}	Ground-to-line voltage in the phase c (V)

1. INTRODUCTION

2015 United Nations Climate Conference (COP21) [1] reinforced that a rapid and global transition to Renewable Energy Sources (RES) offers a realistic means to achieve sustainable development and avoid catastrophic climate change. One key action that must be addressed in order to enable the significant scale-up of RES is to introduce greater flexibility into energy systems and accommodate the variability of these resources [2].

When the intermittent RES share in the generation mix is lower than 15% to 20% of the overall electricity consumption, the Transmission System Operator (TSO) is able to compensate the intermittency. Nevertheless, when the share exceeds 20 - 25%, intermittent RES need to be curtailed during the low consumption periods in order to avoid grid perturbation and grid congestion [3].

The vision for the future Smart Grids includes a significant scale-up of clean energy and energy efficiency that balances environmental and energy goals with impacts on consumer costs and economic productivity. The

adoption of technology for the bidirectional flow of energy and communications would open up access to information, participation, choice, and empower consumers with options from using electric vehicles to producing and selling electricity [4].

In this regard, Energy Storage Systems (ESS) represent a promising key solution to provide ancillary services, essential to facilitate the current and future needs of electricity grids [3, 5–7] playing, therefore, a relevant role in the development of the smart grids [8–11]. Regarding this issue, several studies have been conducted about storage technologies (Table 1), identifying their technical characteristics and most appropriate applications [12–17].

A widely-used approach for classifying EES is the determination according to the form of energy used. In this sense, ESS are classified into mechanical, electrochemical, chemical, electrical and thermal energy [18]. Throughout the supply chain, ESS can be implemented into large-scale energy storage (GW), such as reversible hydro (pumped storage) or thermal storage; storage in grids (MW), like batteries, capacitors and superconducting coils and flywheels; and finally, at an end user level (kW), such as batteries, superconducting coils and flywheels.

In particular, electrochemical ESS offer the flexibility in capacity, sitting, and rapid response required to meet application demands over a much wider range of functions [19], such as grid integration, offering versatility as well as high energy density and efficiency [20] and providing a wide range of services, including voltage control, power flow management, system restoration, energy and ancillary markets, commercial and regulatory framework and grid management [7, 21–26].

Experience so far demonstrates that technological progress is not sufficient to boost storage deployment [27–29]. The integration of grid-connected Battery Energy Storage Systems (BESS) within electrical power systems has been hampered by technology costs, limited deployment experience, existing electricity market and regulatory structures and complex value chains which increase investment risk [9]. This means that technological development should actively complement an adequate regulatory environment, industrial acceptance and progress on different issues still needing regulatory support and research and development funding [29, 30]. In particular, the integration of ESS in grids requires designing topologies and special converters, for virtually each case [21].

Power Electronic Interfaces (PEI) associated to grid-connected BESS are responsible for exchanging power between battery units and loads or the AC-side source [31, 32]. These PEI systems must be bidirectional converters with the capability to operate in both charging and discharging modes [31]. When implementing PEI, one of the major challenges is the design parameters, depending upon the ESS and their implementation.

Voltage source inverters (VSI) are PEI traditionally operated as voltage sources where the controller generates gate pulses for obtaining an output voltage with a particular fundamental magnitude and frequency [33]. There is a range of modulation techniques currently available for controlling three-phase inverters. In this sense, Sinusoidal Pulse Width Modulation (SPWM) is a traditional modulation technique. Within this technique, several modulation algorithms have been proposed, each pretending to improve some characteristic within the process, for example, switching losses, conversion efficiency or harmonic content present in the output wave [34].

Most applications of grid-connected three-phase SPWM VSI provide control of instantaneous current waveform and high accuracy, among other advantages such as peak current protection, overload rejection, extremely good dynamics, compensation of effects due to load parameter changes (resistance and reactance), compensation of the semiconductor voltage drop and dead times of the converter and compensation of the DC-link and AC-side voltage changes. Control structure of this kind of inverters comprises an internal current feedback loop [35], consequently, the inverter efficiency depends largely on the control technique chosen. Hysteresis control is a widely used method for current control because of its simplicity of implementation and fast response current loop [33].

Table 1. Performance of Storage Technologies [16]

Storage Type	Power (MW)	Discharge time	Efficiency (%)	Lifetime (yr)	Overall storage cost (USD/MWh)	Capital cost (USD/kW)
Pumped Hydro	250 – 1000	10h	70 – 80	> 30	50 - 150	2000 –4000 (100 –300)[(b)]
Compressed air energy storage (CAES)	100 – 300 (10/20)	3 – 10h	45 – 60	30	- 150	800–1000 (1300–1800)[(c)]
Fly Wheels	0.1 – 10	15s – 15m	> 85	20	Na	1000 –5000[(d)]
Supercapacitor	10	< 30s	90	5 10^4 cycles	Na	1500 – 2500 (500)[(d)]
Vanadium redox battery (VRB)	0.05-10	2 – 8h	75/80DC 60/70AC	5 - 15	250 - 300[(d)]	3000 – 4000 (2000)[(d)]
Li-ion battery	- 5	15m – 4h	90DC	8 - 15	250 – 500[(d,e)]	2500 – 3000 (< 1000)[(d,e)]
Lead battery	3 - 20	10s – 4h	75/80DC 79/75AC	4 - 8	na	1500–2000
NaS battery	30-35	4h	80/85DC	15	50 – 150[(d)]	100 –2000[(d)]
Superconducting magnetic energy storage (SMES)	0.5+[(d)]	1 – 100s/h[(d)]	> 90	> 5 10^4 cycles	na	na

(a) All figures are intended as typical order of magnitude estimated based on available sources and information, often with wide ranges of variability;

(b) Hydro power plant upgrading for storage service;

(c) Small systems (10 – 20 MW);

(d) Projected/estimated;

(e) Large Li-ion cells

Literature refers to bidirectional converter topologies focused on the operation of direct current (DC) bus distribution systems due to its operation on mini grid implementations [36–41], as well as BESS according to its direct applicability on the grid by the TSO [24, 42–46]. Typical bidirectional converters are based on indirect topologies consist of two or more stages of conversion [47].

Frequent solutions include a double-stage conversion. In [36] a double-stage high-efficiency isolated single-phase bidirectional AC–DC converter for a 380 V DC power distribution system is proposed. The power module is mainly composed by an AC-DC rectifier for grid interface and an isolated bidirectional DC-DC converter to interface DC bus and DC link of the rectifier. The 5 kW prototype developed achieved an overall power conversion efficiency of almost 96% at 2.5 kW and 94.5% at the full load of 5 kW. In [39] an energy management system for DC loads is designed. In this sense a 5 kVA prototype composed by a bidirectional DC-DC converter with an intermediate high-frequency transformer together with a three-phase inverter is developed. The device operates at near-unity power factor values with in an energy management scheme that ensures effective utilization of energy in the DC nanogrid by supplying uninterrupted power while minimizing grid utilization. In [44] a transformerless power inverter scheme is simulated and verified experimentally for a split phase power system 240 V/120 V, reaching an average power conversion efficiency of 95.4% for a total power of 5.3 kW focused on residential battery storage system. In [48] developed a 100 W prototype with low-side voltage of 24 V and high-side voltage of 230 V with an operation mode divided in two stages of power flow and two stages of control. The power module has an isolated bidirectional converter together with a bidirectional inverter designed for supplying power to consumer during peak loads. Even a less common solution carried out in [47], a three-phase bidirectional AC-DC converter is proposed, formed by three single-phase direct AC-DC converters, using a six-leg inverter connected to a three-phase direct AC/AC converter by a high-frequency link transformer. A 20 kW prototype developed shows 94.2% efficiency when the DC port is receiving power from the AC port and 94.5% for the reverse direction. Advantages of the proposed AC-DC converter include the absence of a bank of capacitors, transformer isolation and lower voltage levels across the switches. According the authors, the main drawback of the proposed converter is the large number of switches and drivers and, consequently, lower reliability.

On the other hand, direct conversion topologies have a single-stage to carry out the AC-DC conversion [37, 38, 45, 49–51] that have resulted in increased efficiency, reduced size, high reliability and compactness of bidirectional converters [52]. DC-bus voltage regulation for a 380 ± 10 V DC distribution system is presented in [37] integrated with a 7 kW three-phase single-stage bidirectional inverter, and reassessed in [38] to enhance the voltage ripple in single-phase converters with a 5 kW single-stage bidirectional device. Converters are not technically specified in detail but the regulation approach used allow to reduce the mode-change frequency and the chance of under/over voltage protection and can improve the operational reliability and availability. In [45] a three-level configuration with a storage set in two battery banks connected in series with connection to ground in the mid-point is proposed. Moreover, a theoretical modulation technique and the control algorithm associated with the correct AC signal generation has been proposed, reaching power factor values close to unity. In [49] a 3.7 kW single-phase, single-stage, bidirectional and isolated AC-DC converter prototype is developed which interfaces a 400 V DC-bus with the utility grid. Experimental results show high conversion efficiencies over 96% and power factor values close to unity. In [50] a single-stage, single-phase, three-input AC-DC converter is proposed, which interfaces two unidirectional input ports for photovoltaic and fuel cell sources and two bidirectional ports for an ultra-capacitor and a resistive load. The proposed converter does not need any output voltage filter minimizing passive elements, utilizes minimum number of power switches (only six) achieving efficiency average values over 90%. The reduction of elements in the design of devices has resulted in promising structures such as [51], where a 1 kVA four-switch bidirectional converter prototype, connected in an open delta configuration to a line-to-line rms voltage of 120 V is developed. Although, this cost effective structure provides energy saves by reducing conduction losses, it also suffers increased voltage stress on the power devices, increased switching losses in each device due to high DC bus voltage and fluctuation of voltage across individual split capacitors.

As mentioned above several prototypes of bidirectional converters have been proposed and effectively presented in literature [36–39, 44, 45, 47–51] covering not only single-phase [36, 38, 44, 48–50], but also three-phase topologies [37, 39, 45, 47, 51]. Nevertheless the prototypes has been developed as power units up to 20 kW [36–39, 44, 45, 47–51] and just a few studies have been conducted within the framework of three-phase grid connected bidirectional converters [45, 47, 51].

This study designs, implements and evaluates the operation of a single-stage three-phase SPWM VSI, consisting of two modules of 150 kW, for operating as a grid connected AC-DC bidirectional converter, in such a way that has the ability to operate in charging mode storing electrical energy, or in discharging mode supplying power depending on demand. The design has been carried out with MATLAB Simulink [53] environment and the performance has been complemented by a sensitivity analysis of the operating parameter set by the TSO. In addition, the AC-DC bidirectional converter has been implemented in a controlled environment in order to study the behaviour of grid-connected BESS under real operating conditions.

After this introduction section, Section 2 presents the BESS configuration as well as the bidirectional converter operating principle including the simulation proposed for the charge/discharge function. Simulation is complemented with a sensitivity analysis to evaluate how the parameters set by the TSO influence the model output. Section 3 details the implemented system describing its components, to continue in Section 4 with the results and analysis grid connected test and in Section 5 with the discussion. Finally, Section 6 concludes the study summarizing the main results.

2. System Configuration and Modelling

2.1. Design of the Three-Phase Hysteresis Controlled VSI Converter

The design allows the use of a single hardware (converter) in two operation modes enabling the bidirectional energy flow. Discharge mode involves the extraction of energy from the battery bank to the grid. This mode requires synchronizing the output current of the bidirectional converter to grid voltage in order to approximate the power factor to 1 and thereby minimize the reactive power. Alternatively, charge mode uses the grid to recharge the battery bank and store energy. This is achieved by rectifying the grid voltage and regulating the current flow in the batteries. Figure 1 shows the system block diagram which comprises an AC-DC bidirectional converter connected to the AC grid and a DC battery bank. The AC-DC bidirectional converter make the conversion by a power module whose signal is conditioned by an LC filter on the AC-side and by a DC-link composed of a capacitor bank in the DC-side.

The main circuit schematic is represented in Figure 2. The storage is composed by two lead-acid battery banks connected in series, with the neutral connected to the DC midpoint. The use of lead-acid batteries is widespread in grid-connected applications because of its robustness and reliability, besides the low self-discharge rate, fast response times, relatively high cycle efficiencies and low capital costs [12, 14]. In any case, several electrochemical technologies are considered during simulations, in order to test the system accuracy. Moreover, a DC-link filter composed by capacitors is connected between the battery bank and the power module in order to fix the voltage. The power module consists of three Insulated Gate Bipolar Transistors (IGBT) semi-bridges, commanded by an activation signal, so that each semi-bridge generates a phase [54]. The LC-filter located at the power module output filters all high frequency components caused by transistors switching. Voltage on the AC-side is fixed by the grid, therefore it is necessary to implement an appropriate current control strategy in the operation of the bidirectional converter [35].

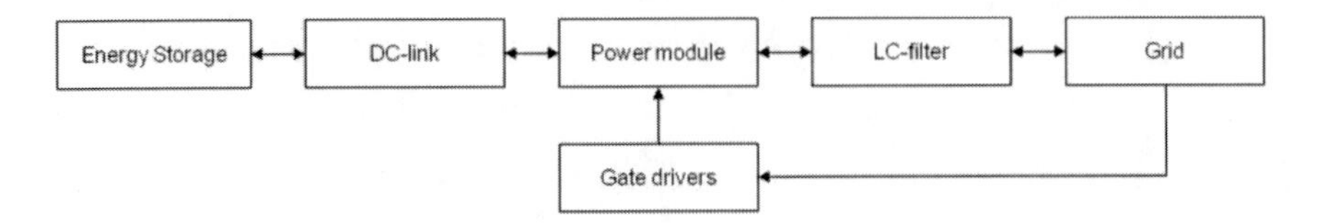

Figure 1. System block diagram.

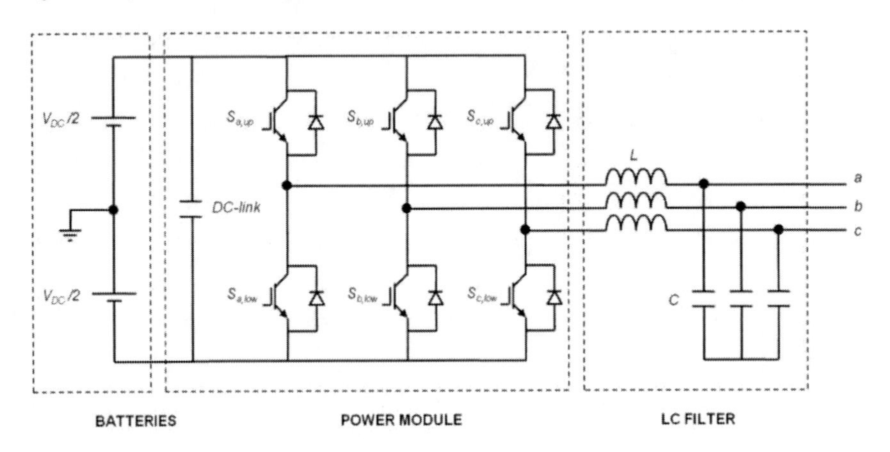

Figure 2. Main circuit schematics.

2.2. Working Principle: Fixed-Band Hysteresis Controller

Current controlled SPWM inverters are widely used in high-performance alternating current (AC) drives. The main objective of the current controller is to force the load current vector according to reference current trajectory. In current controller, load currents are measured and compared with reference currents, and the errors are used as an input to the SPWM modulator, which provides inverter switching signals [55].

Among all current control techniques for SPWM, the hysteresis current control is widely used because of its simplicity of implementation; fast response [35, 55–57]. Hysteresis current control is a non-linear control technique [58] extensively used in high performance AC systems because provides high dynamic response besides the simplicity of implementation [55]. The converter switching frequency depends largely on the load parameters and the hysteresis bandwidth, and varies with the AC voltage [35].

The hysteresis band can be programmed as a function of the load and supply parameters in order to maintain a fixed modulation frequency [57]. The fixed-band hysteresis controller gives good performance except that the switching frequency is irregular and current ripple is relatively large [58]. A low pass filter, composed by an inductor and a capacitor set in parallel, is used to minimize the ripple and, hence, the high frequency harmonic content [57].

Hysteresis controller contains a nonlinear feedback loop with hysteresis comparator. Thought the hysteresis comparator controller the steady-state error could be controlled by varying the hysteresis band (HB) width [58, 59].

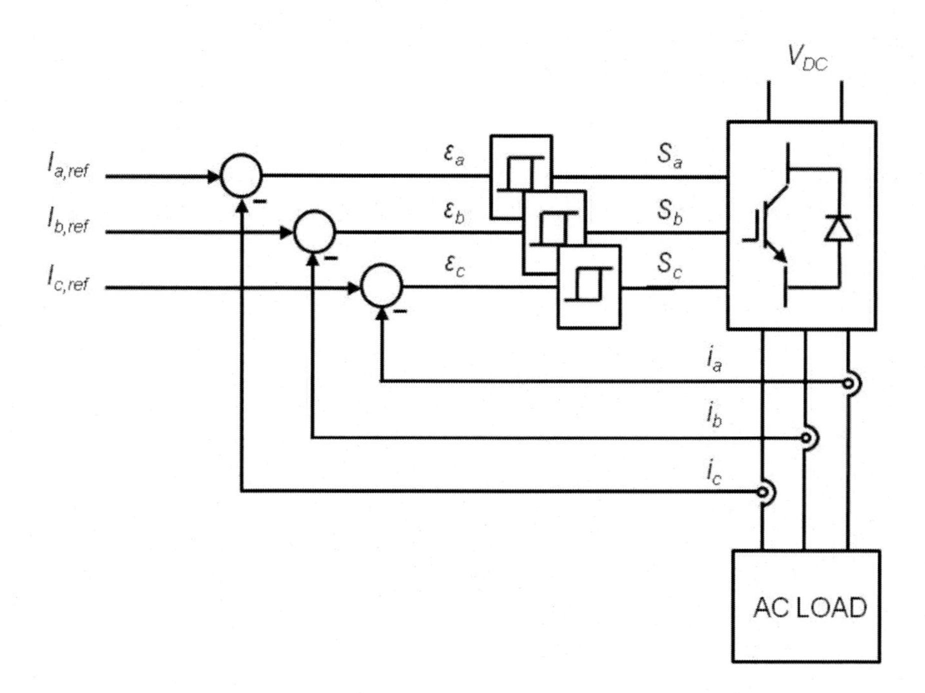

Figure 3. Hysteresis current controller diagram for a three-phase VSI.

The hysteresis comparator has two bands, a lower hysteresis band (LHB) and an upper hysteresis band (UHB) which determines the steady-state error. In Figure 3, sinusoidal reference currents ($i_{a,ref}$, $i_{b,ref}$, $i_{c,ref}$) are compared with the respective measured phase currents (i_a, i_b, i_c), resulting error signals (ε_a, ε_b, ε_c), which command, thought the HB current controller, the transistor base drives (S_a, S_b, S_c).

The concept of the voltage (or current) vector is utilized because it is a very convenient representation of a set of three-phase voltages (or currents). The voltage vector is defined by the following expression [59]:

$$\bar{v} = \frac{2}{3}(v_a + \bar{a}v_b + \bar{a}^2 v_c) \quad (1)$$

where $\bar{a} = e^{j(2\pi/3)}$ defines a two-dimensional vector (or complex number) associated with the three-phase voltages (v_a, v_b, v_c).

The actual voltages can be recovered from voltage vector, $\bar{v}$, and the zero sequence component, v_0, using the following equations:

$$v_a = |\bar{v}| \cdot cos\theta + v_0$$
$$v_b = |\bar{v}| \cdot \cos\left(\theta - \frac{2\pi}{3}\right) + v_0$$
$$v_c = |\bar{v}| \cdot \cos\left(\theta + \frac{2\pi}{3}\right) + v_0$$

where θ is the angle between the voltage vector and real axis.

The bidirectional converter operates in one of eight conduction modes to produce one of six nonzero voltage vectors or a zero voltage vector as illustrated in Figure 4 [35]. Due to the neutral is connected to the DC bus midpoint between the two battery banks (Figure 2), the line-to-ground voltages (v_{ag}, v_{bg}, v_{cg}) are equal to the line-to-neutral voltages ($+v_{DC}/2, -v_{DC}/2$) at the DC-side [59].

Similarly, the current vector is defined as:

$$\bar{\imath} = \frac{2}{3}(i_a + \bar{a}i_b + \bar{a}^2 i_c) \quad (2)$$

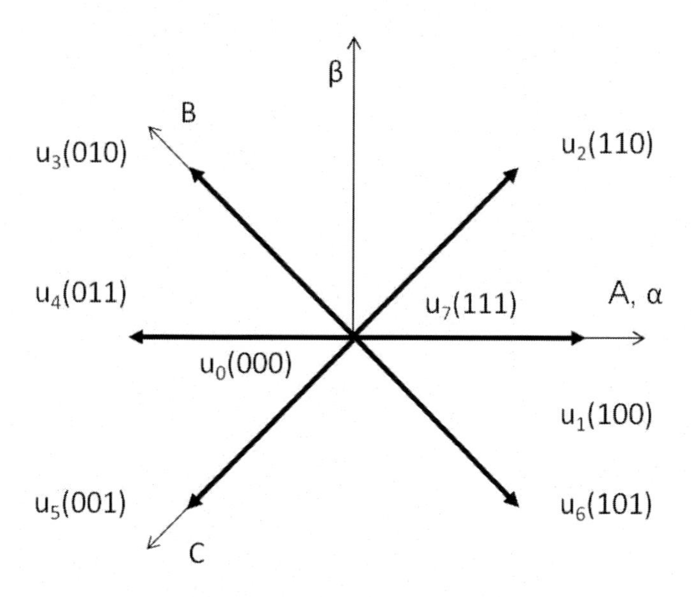

Figure 4. Voltage vector of VSI converter [35].

Taking as example the phase a(becomes equivalently for the other phases), the algorithm of this fixed band control scheme (Figure 5) is given by [57]:

$$i_{a,ref} = I_{max}\, sin(\omega t)$$

$$i_{a,up} = i_{a,ref} + h$$

$$i_{a,low} = i_{a,ref} - h$$

$$If\ i_a > i_{a,up};\ v_a = -\frac{v_{DC}}{2}$$
$$If\ i_a < i_{a,low};\ v_a = +\frac{v_{DC}}{2}$$

where $i_{a,up}$ is the *UHB* of the phase a, $i_{a,low}$ is the *LHB* of the phase a, I_{max} is the peak current value, ωt is the angular velocity per unit time, *h* is the hysteresis band limit and v_{DC}. is the terminal voltage of the battery bank.

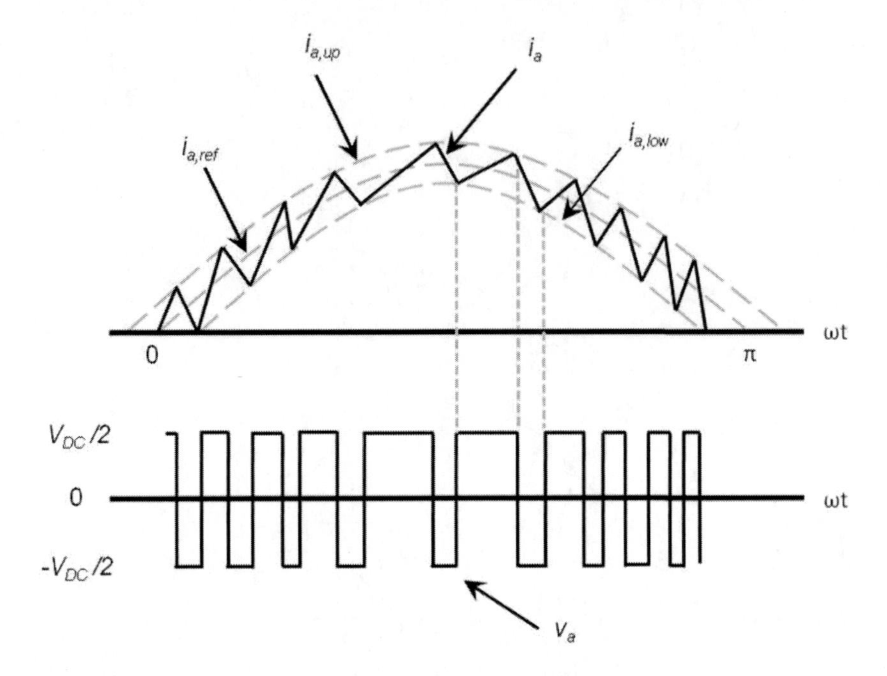

Figure 5. Band shape of the fixed-band hysteresis controller.

From Figure 5, if $i_a > i_{a,up}$ the output in the hysteresis module is zero, which means that the bidirectional converter output voltage is negative in order to reduce the line current. In the same way, if $i_a < i_{a,low}$ the output in the hysteresis module is one, which means that the bidirectional converter output voltage is positive in order to increase the line current [55].

2.3. Simulation of the Three-Phase Hysteresis Controlled Inverter

In power electronic systems, simulation is performed essentially to control and design the circuit configuration and the control strategy to be applied [60]. The simulation has been carried out with MATLAB® Simulink® environment [53] assessing the performance of the bidirectional converter during load variation.

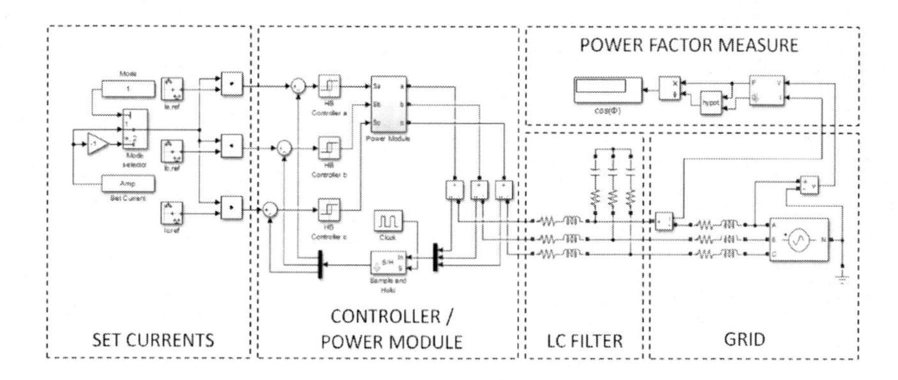

Figure 6. Model scheme for the three-phase bidirectional converter designed under MATLAB® Simulink® environment.

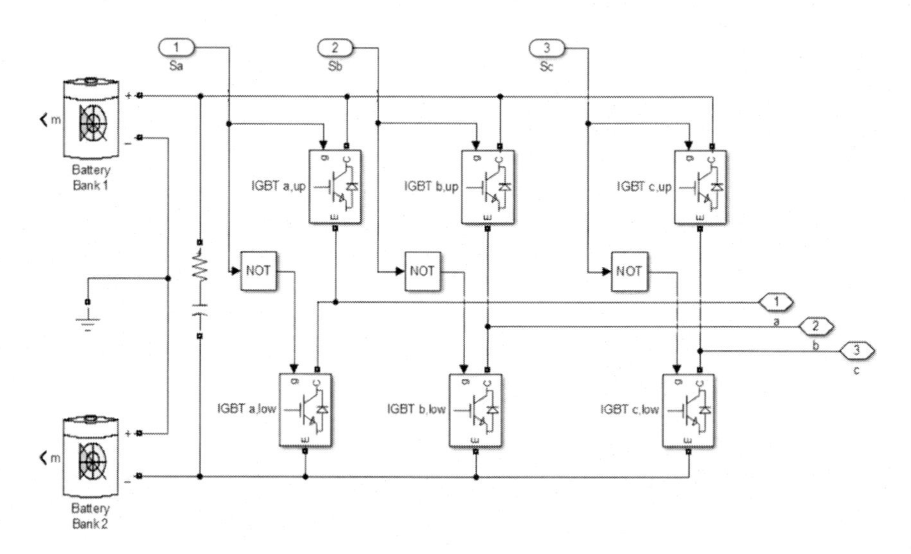

Figure 7. Power Module subsystem included in the model scheme.

In order to simulate the bidirectional operation of a three-phase VSI converter, a model is implemented following the operational structure previously described. In Figure 6 the resulting model scheme is depicted. The current value and the operation mode are set in order to define the reference currents and the direction of the current flow in the Controller/Power Module stage. LC filter is connected to the power grid, where the current and voltage are measured in order to obtain the power factor, cos(ϕ), so that the optimal operation of the bidirectional converter can be assessed. Likewise, Figure 7 displays the power module subsystem where the IGBT semi-bridges and the battery banks are included.

Table 2 shows the main parameters of the simulation performed, which match with the parameters used in the implementation of the bidirectional converter described in Section 3.

Table 2. BESS parameters

Circuit parameters	
V_{DC}	720 V
I_{ref}	150 A_{rms} per phase
Capacitor DC-link	3 x 420 μF in parallel
Parasitic resistance of capacitor DC-link	3 x 2.3 mΩ in parallel
Inductance L	2.3 mH per winding
Parasitic resistance of inductance L	5 mΩ per winding
Capacitor C	2 x 50 μF per phase
Parasitic resistance of capacitor C	2 x 12 mΩ per phase
Inductance L_{grid}	0.23 mH per phase
Parasitic resistance of inductance L_{grid}	82 mΩ per phase

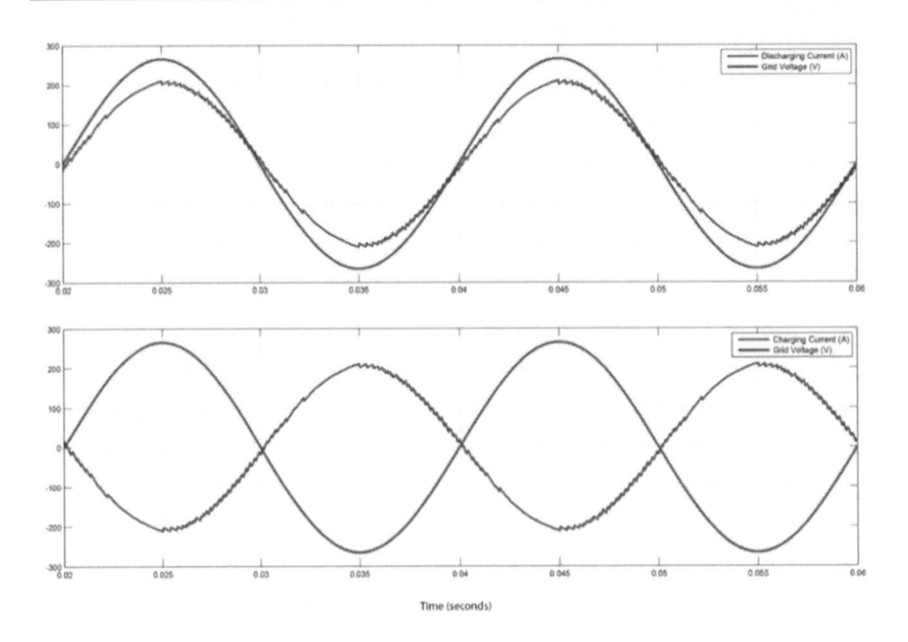

Figure 8. Simulated operations for the VSI converter in discharge mode (up) and charge mode (down).

Each phase is associated with a reference current signal, 120º degrees out of phase, where each amplitude is defined depending on the charge and discharge requirements of the battery banks. Each reference signal is defined using a finite sample of 90 values, being the sampling frequency of 4,500 Hz, to achieve a 50 Hz frequency set by the grid. Charge/discharge function is determined from the sign of the reference signal. In this way, as shown in Figure 8, if the reference signal is in phase with thc grid signal, the bidirectional converter operates in discharge mode. However, if the reference signal is offset 180° from the grid signal, the bidirectional converter operates in charge mode.

Reference currents are compared with the measured currents obtained in the AC-side of the bidirectional converter. In order to accommodate the measured grid current, a sample-and-hold block is used to limit the maximum switching frequency of the bidirectional converter to 30 kHz.

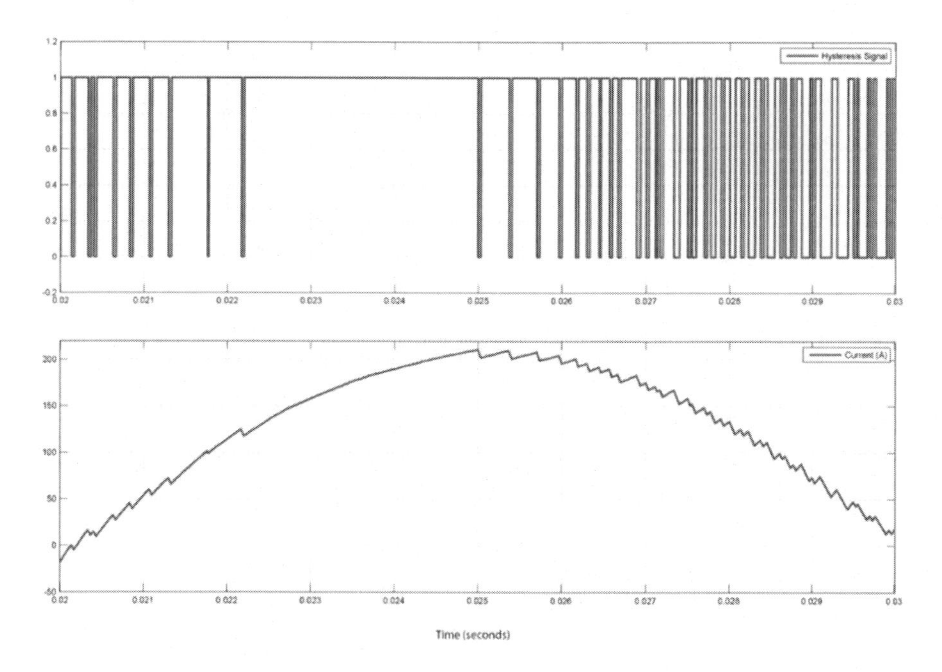

Figure 9. Simulated hysteresis signal.

From that comparison, error signals are obtained, through hysteresis module, commanding the IGBT semi-bridges Figure 9. When measured current signal exceeds the upper value of the HB, the output in the hysteresis module is zero, producing a voltage decrease and, consequently, a current decrease (Figure 9). Similarly, when measured current signal is below the lower value of the HB, the output in the hysteresis module is one, producing a voltage increase and, consequently, a current increase. This behaviour is based on the principle of operation of the inductor of the LC Filter and how the voltage drop depends on the value of the inductance and the variation of the current passing through it. Therefore, the hysteresis pulse width depends on the voltage drop and the time period in which the current that passes through the coil exceeds the upper HB in the positive cycles of the generated signal (Figure 9 below). Similarly, the same behaviour is obtained for the lower HB values in the negative cycles of the generated signal.

2.4. Sensitivity Analysis

Connectivity and operation capacity of the BESS can be attributed to variations in the grid signal. Therefore, it is necessary to evaluate the impact in the output of the model to determine the operational ranges of the bidirectional converter. In this regard, a sensitivity analysis allows exploring the impact on the output of the model from the variation of key input parameters [61–63].

In this particular case, the model output is affected, basically by changes in reference currents set by the TSO. The power factor, $\cos(\phi)$, is a key output quality parameter in order to evaluate the optimal operation of the bidirectional converter [64].

A Global Sensitivity Analysis has been performed using the Sensitivity Analysis Tool® from MATLAB® Simulink® environment [53]. In order to evaluate how changes in the reference current influence the output requirements, a parameter sample space has been defined considering 16A increments in the reference current. Simulations have been carried out through Monte Carlo method, establishing a power factor value over 0.99 as an optimal design requirement.

Considering that the voltage is 230/400 V, the analysis is performed for reference currents in a range 15 A to 271 A, equivalent to a power range between 10 kW and 188 kW, as the assessed input parameter. Regarding the system power rating developed in this research, and according to the guidelines established by the Electric Power Research Institute (EPRI) in terms of application requirements and technology characteristics for different energy storage technology options [65], the analysis has been performed by simulating three electrochemical energy storage technologies such as Lead-acid (Pb-Acid), Lithium Ion (Li-ion) and Nickel-Metal Hydride (Ni-MH).

Figure 10 shows the results from Global Sensitivity Analysis. For the three technologies evaluated, the values of power factor over the design requirement set as 0.99 are obtained for reference currents greater than 79 A, equivalent to 55 kW or approximately 1/3 of the bidirectional converter rated power. The greater frequency of points are likewise in the described range, and it follows that the optimum operating range of the bidirectional converter corresponds to reference current values higher than 79 A. These results reflect the robustness of the three-phase bidirectional converter regardless the electrochemical energy storage system selected.

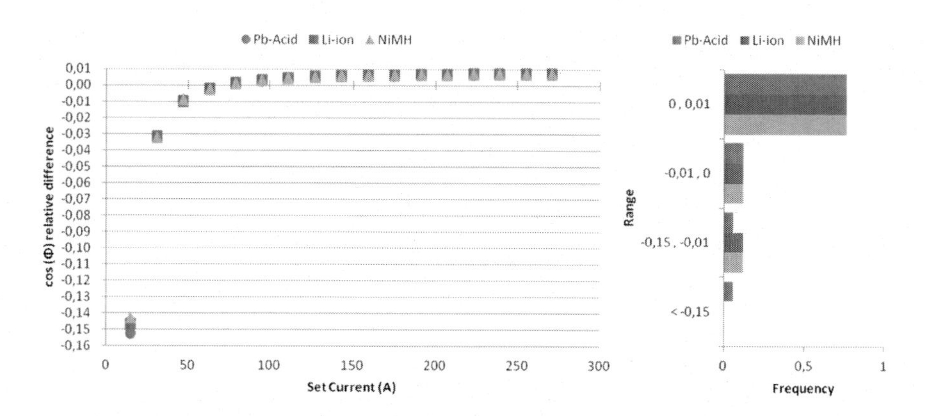

Figure 10. Results from the global sensitivity analysis performed with the Sensitivity Analysis Tool® from MATLAB® Simulink® [53] environment.

3. Laboratory Implementation

The operation of the ESS proposed is based upon a bidirectional converter which undertakes a battery charging/discharging function as from reference current signals. For convenience, systems are mounted in modules of two bidirectional converters of 150 kW each (Figure 11). If necessary, this modular design allows a scale up of the system according to the increase in power. Therefore, the storage and energy-flow capacities would be modified by adjusting the number of batteries and converters according to requirements, which increases the system reliability as well as the feasibility and simplicity in operation and maintenance.

Figure 11. Module composed by two bidirectional converter of 150kW each manufactured by ITER, S.A.

The bidirectional converter is mainly composed by the power module, the LC filter and the control board. A power source has been designed and implemented to supply the control board and the power module, including a line filter in order to avoid the noise at the AC input of the source, as well as a varistor and fuses as protections against overvoltage and overcurrent.

The whole system is completed with 360 Lead-acid batteries, Ecosafe/ Hawker® model TVS7 (Figure 12), set in two battery banks in series and connected to ground in the mid-point, where v_{DC} reaches a total amount of 720 V. The total capacity of the assembly is about 526 kWh with a measured internal resistance of 0.3 Ω.

3.1. Power Module

The semiconductor power stage is formed by three IGBT semi-bridges so that each semi-bridge generates a phase. The power module selected is a Semikron® Skiip 613 GD123 (Figure 13). The Skiip module has temperature, current and voltage sensors as well as protections against shortcircuit, overtemperature, overcurrent, overvoltage, etc. Likewise, several command and alarm signals are available.

Figure 12. Lead-acid battery storage system installed at ITER, S.A.

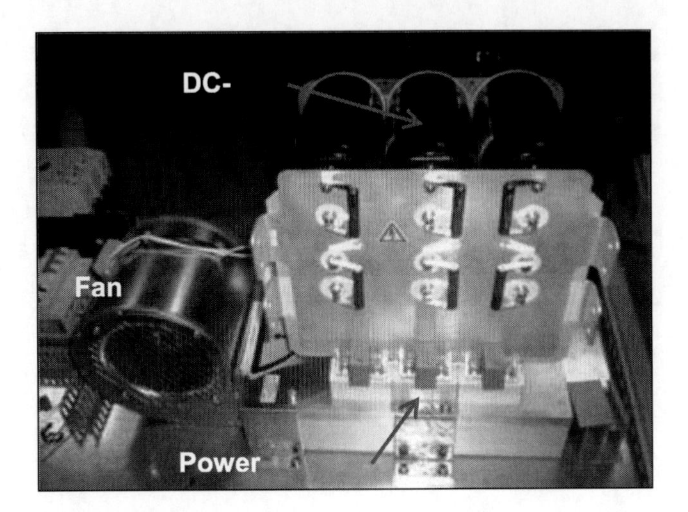

Figure 13. View of the Semikron® Skiip 613 GD123 power module.

The DC-link filter is designed to cover a wide range of input voltages. It consists of a bank of three polypropylene capacitors 420 uF 1100 V located in the DC input to reduce overvoltage peaks that occur during switching of the IGBTs.

3.2. LC Filter

The LC filter consists in a sinusoidal filter conformed in T and composed by a three-phase inductor of 2.3 mH per phase and 2 capacitors 50 μF 400 V per phase. It is designed to filter the high frequency components generated as from the high switching frequencies of the IGBT power module. In this sense, the filter should have the less resistance to the generated 50 Hz signal.

3.3. Control Board

The control stage receives via software the reference currents defined in magnitude and direction in order to set the batteries charge/discharge function. In the implemented solution the control board compares directly the measured currents with the reference currents instead of generate an error signal as mentioned before. Moreover, the hysteresis band limit is set as zero. In consequence, when the value of the measured current at the output of the bidirectional converter exceeds the reference current, the lower IGBT are commanded to switch and the upper IGBT to stop what implies decreasing the output current. Similarly, when the measured current falls behind the reference current, the upper IGBT are commanded to switch and the lower IGBT stop increasing the output current. In order to reach the 50 Hz fixed by the grid, the reference signal is defined, as mentioned above, using a finite sample of 90 values being the sampling frequency of 4,500 Hz. The control scheme is displayed in Figure 14.

Data transmission is performed via MODBUS RTU (Remote Terminal Unit) communication protocol for RS485 serial port on a Supervisory Control and Data Acquisition (SCADA) system. The monitoring and alarm detection firmware have been implemented to minimize the response time in case of any fault. The control board is designed and manufactured with master/slave architecture with two Microchip® 18F4525 PIC® microcontrollers.

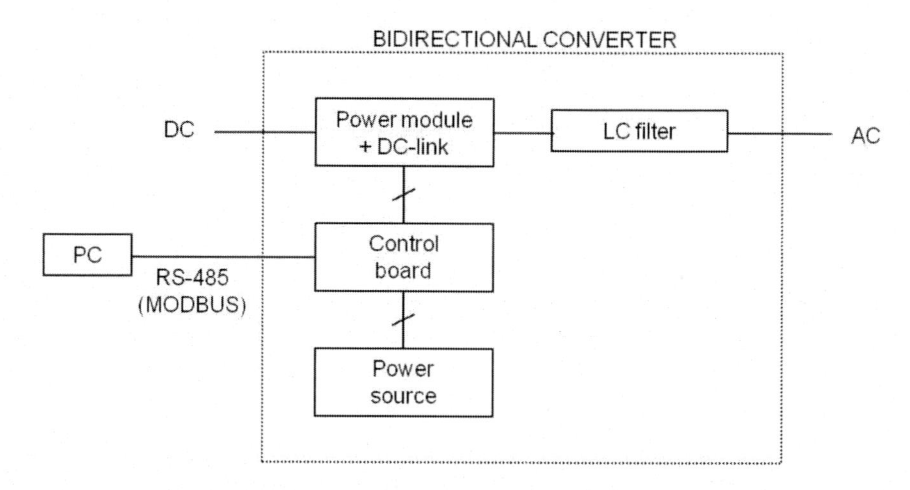

Figure 14. Control scheme of the bidirectional converter.

4. Results and Analysis

The implemented system was connected to an internal AC network (230/400 V_{rms}, 50 Hz). The power and performance measures have been carried out with the Power Analyzer Zes Zimmer® LMG500. This is a precision instrument that has dual channels for simultaneously current and voltage data acquisition. The measurement accuracy for DC parameters is 0.02% for current and voltage and 0.032% for power.

Reference currents are introduced in the power module through the control board, therefore, since the voltage is fixed by the grid, the references will set an objective power in the AC side. Measurements have been carried out considering a wide range of reference currents from 31 A to 255 A, equivalent to a power range from 21 kW to 177 kW. As energy is flowing to the grid, the batteries are discharged so that the terminal voltage decreases. In this sense, in order to maintain the value of fixed power facing the DC-AC conversion, the injection of current from the batteries is increased slightly (Figure 15). Likewise, for higher reference currents, over 159 A, the depth of discharge of the batteries is reduced.

In Figure 16 the efficiency curves for the bidirectional converter in the discharge mode for different reference currents depending on the battery voltage are shown. Decreasing battery voltage implies a slight increase in the discharging current injection, resulting in a slight increase in conversion efficiency, measured as the ratio of the power in the AC side and the power in the DC side. It is observed that the conversion efficiency for the bidirectional converter is constant for reference currents exceeding 79 A, achieving efficiencies over 98%.

Table 3 shows the measures obtained for the power factor from the different reference currents and battery voltages. It is observed that power factor values over 0.99 are achieved for reference currents exceeding 79 A (grey-shaded area), confirming the resulting values from the global sensitivity analysis undertaken in Section 2.4.

Table 4 summarizes the power measures obtained from the different reference currents and battery voltages. It follows that each bidirectional converter unit must work over 50 kW (grey-shaded area), or what is the same, over 30% of the rated power. For optimal operating conditions, an upper limit of 155 kW is set.

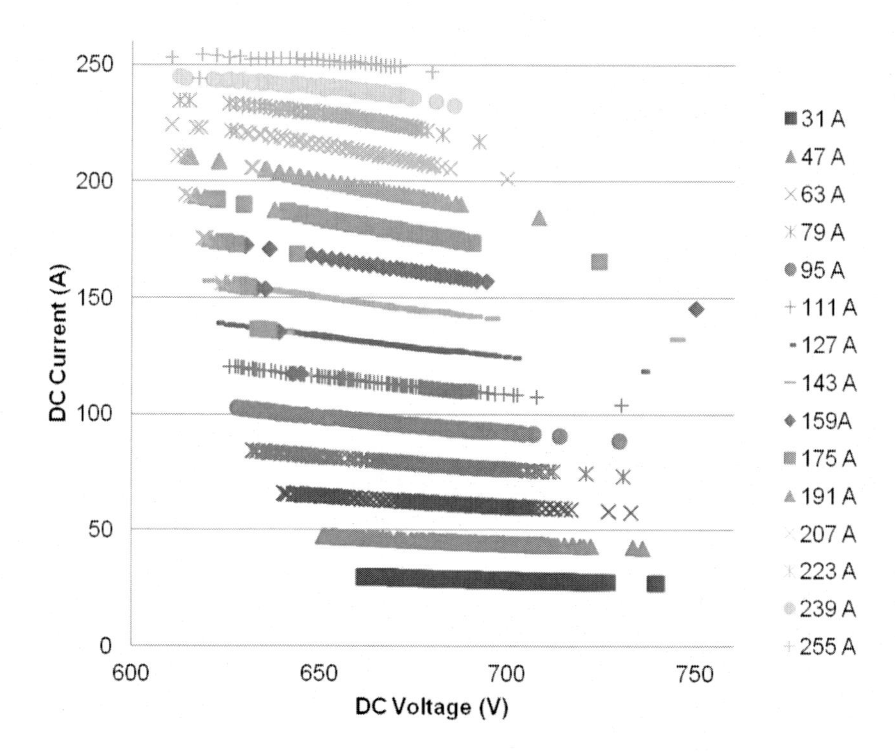

Figure 15. Battery discharge measured data for the different reference currents.

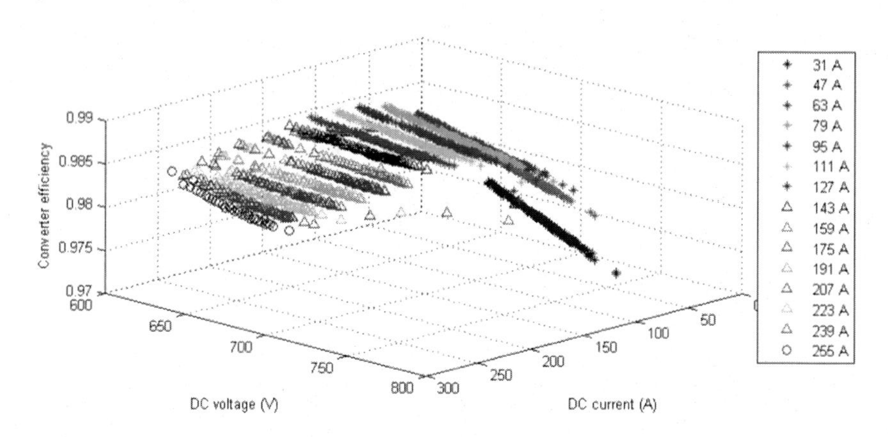

Figure 16. Efficiency for the bidirectional converter in the discharge mode calculated from measured data.

Table 3. Power factor measured according to reference currents and battery voltages

		Reference Current (A)														
		31	47	63	79	95	111	127	143	159	175	191	207	223	239	255
Battery voltage (V)	**720**	0.943	0.975	0.986	**0.990**	**0.994**										
	710	0.943	0.976	0.987	**0.993**	**0.993**	**0.995**									
	700	0.944	0.976	0.987	**0.991**	**0.993**	**0.995**	**0.996**	**0.997**				**0.996**		**0.995**	
	690	0.946	0.976	0.987	**0.991**	**0.994**	**0.995**	**0.996**	**0.997**	**0.997**	**0.997**	**0.996**	**0.996**	**0.996**	**0.995**	**0.994**
	680	0.943	0.978	0.988	**0.992**	**0.993**	**0.996**	**0.996**	**0.997**	**0.996**	**0.997**	**0.997**	**0.997**	**0.996**	**0.995**	**0.994**
	670	0.948	0.977	0.988	**0.992**	**0.994**	**0.995**	**0.996**	**0.997**	**0.997**	**0.997**	**0.997**	**0.997**	**0.996**	**0.996**	**0.994**
	660	0.949	0.977	0.986	**0.992**	**0.994**	**0.995**	**0.996**	**0.996**	**0.997**	**0.997**	**0.997**	**0.997**	**0.997**	**0.996**	**0.994**
	650		0.975	0.987	**0.992**	**0.995**	**0.996**	**0.996**	**0.997**	**0.997**	**0.998**	**0.998**	**0.997**	**0.997**	**0.996**	**0.995**
	640			0.987	**0.992**	**0.994**	**0.995**	**0.996**	**0.997**	**0.998**	**0.998**	**0.998**	**0.997**	**0.997**	**0.996**	**0.994**
	630				**0.991**	**0.994**	**0.996**	**0.996**	**0.997**	**0.997**	**0.997**	**0.997**	**0.997**	**0.997**	**0.995**	**0.995**
	620						**0.995**	**0.996**	**0.997**		**0.998**	**0.998**	**0.998**	**0.997**	**0.995**	**0.994**
	610											**0.998**	**0.998**	**0.997**	**0.996**	**0.994**

Table 4. Discharging AC power (kW) measured according to reference currents and battery voltages

		Reference Current (A)														
		31	47	63	79	95	111	127	143	159	175	191	207	223	239	255
Battery voltage (V)	720	19.4	30.4	41.4	52.5	63.5										
	710	19.3	30.3	41.4	52.4	63.5	74.5									
	700	19.3	30.2	41.2	52.4	63.5	74.5	85.6	96.6				137.6		155.3	
	690	19.3	30.2	41.2	52.2	63.3	74.4	85.6	96.5	107.0	117.7	128.0	137.6	146.8	155.3	163.5
	680	19.2	30.2	41.2	52.2	63.2	74.1	85.3	96.5	106.9	117.7	128.0	137.7	146.7	154.9	163.0
	670	19.3	30.1	41.2	52.2	63.1	74.3	85.2	95.9	107.0	117.9	127.9	137.8	146.9	153.9	161.6
	660	19.2	30.3	41.2	52.2	63.2	74.2	85.2	96.1	106.8	117.6	128.0	137.8	146.3	153.1	160.0
	650		30.3	41.3	52.2	63.2	74.3	85.2	96.0	106.9	117.7	127.7	137.4	145.6	151.4	158.1
	640			41.3	52.3	63.4	74.2	85.2	96.1	106.9	117.8	127.9	137.1	144.8	149.7	155.5
	630				52.4	63.4	74.3	85.2	96.1	106.8	117.5	96.1	136.4	143.6	148.0	155.7
	620						74.3	85.3	96.0		117.4	117.4	135.4	141.4	146.4	153.6
	610											117.4	134.5	140.8	146.8	150.9

5. DISCUSSION

Throughout this paper, the system configuration and model proposed, the simulations carried out and the laboratory implementation results show that the proposed single stage three-phase voltage source bidirectional AC-DC converter has a reliable performance for grid connected bidirectional energy flows in BESS applications.

The operation of the BESS bidirectional design is performed by a six IGBTs power module, a LC filter on the AC-side that is connected to the utility grid and a DC-link composed of a capacitor bank in the DC-side that is connected to a bank of batteries. As mentioned before, the converter is three-phase and single-stage, in such a way the conversion is direct removing the use of intermediate devices as DC-DC or AC-AC converters. Hence, the reduced number of power devices cut down complexity and also results in a decrease of conduction losses.

System configuration and control strategy applied have been properly simulated with MATLAB® Simulink® environment in order to assess the operation of the three-phase converter. Simulation results effectively show the proper bidirectional operation according to set currents and the current flow direction, presenting both modes a low ripple and absolute values of cos(ϕ) close to unity. Therefore, the high dynamic response besides simplicity of implementation of the current control technique used has been demonstrated. Moreover, the assessment is complemented by a global sensitivity analysis varying the operating parameters set by the TSO for three electrochemical energy storage technologies, Pb-Acid, Li-ion and Ni-MH. Results reflect the robustness of the three-phase bidirectional converter regardless the electrochemical energy storage system selected, being the optimum operating range for reference current values higher than 79 A (equivalent to 55 kW)

In order to test the grid connected operation, a BESS composed by a 526 kWh bank of lead-acid batteries and a 150 kW bidirectional converter, is developed. A current sweep from 31 A to 255 A has been performed showing that the power factor for the bidirectional converter exceeds values over 0.99 for set currents higher than 79 A, which matches with the simulation results. In this sense, two important features are exposed compared to the converter referenced in the literature that is the larger operational rated power of 150 kW and the wide range of convertor loads over the 30% of the bidirectional converter rated power, achieving high efficiencies over 98%. The current control technique proposed has been validated enabling a dynamic conversion response depending on power grid fluctuations. Likewise, the control board developed facilitates monitoring and alarm detection via SCADA minimizing faults during BESS operation.

The bidirectional AC-DC converter developed represents a promising solution for the integration of ESS into distribution grids by facilitating the future development of smart grids. Its strengths lie in the high efficiency bidirectional operation feature, as well as the robust design of an integrated system conformed as a single-stage device for grid connection, and expandable to higher powers due to its modular characteristic. This advanced technology represents a low-cost option for proper control and integration of new interconnected storage systems, while maintaining the quality of service, resiliency, and reliability required by TSO and providing the ability to reliably and affordably management of variable energy resources.

CONCLUSION

The present research presents the design and implementation of a BESS operated by a three-phase voltage source bidirectional AC DC converter which undertake a battery charging/discharging function enabling bidirectional energy-flow. Results obtained from laboratory implementation coincide with the operational ranges defined in the simulation and sensitivity analysis, establishing a wide range of optimal operation from 50 kW to 155 kW with power factor values over 0.99 and efficiency values over 98%. This confirms that the implemented current control based on a fixed band hysteresis control provides satisfactory results for the different reference values. Besides the reference currents, the control board establishes the energy-flow direction through the power module based upon IGBT technology, from the out phase between the reference and grid signals. In this sense, if the reference and grid signals are in phase, the bidirectional converter operates as an inverter. Similarly, if the reference and grid signals are 180º out phase, the bidirectional converter operates as a charger.

The bidirectional AC-DC converter developed represents a robust and energy-efficient option for proper control and grid integration of storage systems, while maintaining the quality of service, resiliency, and reliability required by TSO. The operation capacity guarantees the security and quality of the management variable energy resources.

For future implementations, based on the system principle of modularity, it will be required the development of a control that would allow converters to work together in the optimum operational point. For example, for two converters and values over the rated power, it would be preferable that the converters come into operation in an aggregated manner, covering each the 50% of the demand.

REFERENCES

[1] Climate Action. *United Nations Conference on Climate Change (COP21) 2016*. http://www.cop21.gouv.fr/en/ (accessed 6 January 2016).

[2] International Renewable Energy Agency (IRENA). REmap: Roadmap for a Renewable Energy Future. *IRENA*, 2016.

[3] European Commission - DG Energy. *The future role and challenges of Energy Storage.*, 2013.

[4] U.S. Department of Energy. *Grid Energy Storage*, 2013, 67.

[5] EAC. *2014 Storage Plan Assessment: Recommendations for the US Department of Energy*, 2014.

[6] Pearre, NS; Swan, LG. Technoeconomic feasibility of grid storage: Mapping electrical services and energy storage technologies. *Applied Energy*, 2015, 137, 501–10. doi:10.1016/j.apenergy.2014.04.050.

[7] Kondoh, J; Ishii, I; Yamaguchi, H; Murata, A; Otani, K; Sakuta, K; et al. Electrical energy storage systems for energy networks. *Energy Conversion and Management*, 2000, 41, 1863–74.

[8] Eyer, J; Corey, G. *Energy Storage for the Electricity Grid*: Benefits and Market Potential Assessment Guide A Study for the DOE Energy Storage Systems Program. Sandia National Laboratories, 2010.

[9] Lyons, PF; Wade, NS; Jiang, T; Taylor, PC; Hashiesh, F; Michel, M; et al. Design and analysis of electrical energy storage demonstration projects on UK distribution networks. *Applied Energy*, 2015, 137, 677–91. doi:10.1016/j.apenergy.2014.09.027.

[10] Subburaj, AS; Pushpakaran, BN; Bayne, SB. Overview of grid connected renewable energy based battery projects in USA. *Renewable and Sustainable Energy Reviews*, 2015, 45, 219–34. doi:10.1016/j.rser.2015.01.052.

[11] Oliveira, L; Messagie, M; Mertens, J; Laget, H; Coosemans, T; Van Mierlo, J. Environmental performance of electricity storage systems for grid applications, a life cycle approach. *Energy Conversion and Management*, 2015, 101, 326–35. doi:10.1016/j.enconman.2015.05.063.

[12] Luo, X; Wang, J; Dooner, M; Clarke, J. Overview of current development in electrical energy storage technologies and the application potential in power system operation. *Applied Energy*, 2015, 137, 511–36. doi:10.1016/j.apenergy.2014.09.081.

[13] Cho, J; Jeong, S; Kim, Y. Commercial and research battery technologies for electrical energy storage applications. *Progress in Energy and Combustion Science*, 2015, 48, 84–101. doi:10.1016/j.pecs.2015.01.002.

[14] Raza, SS; Janajreh, I; Ghenai, C. Sustainability index approach as a selection criteria for energy storage system of an intermittent renewable energy source. *Applied Energy*, 2014, 136, 909–20. doi:10.1016/j.apenergy.2014.04.080.

[15] Zakeri, B; Syri, S. Electrical energy storage systems: A comparative life cycle cost analysis. *Renewable and Sustainable Energy Reviews*, 2015, 42, 569–96. doi:10.1016/j.rser.2014.10.011.

[16] IRENA, IEA-ETSAP. Electricity Storage. *Technology Brief.*, 2012, 28.

[17] Ciez, RE; Whitacre, JF. Comparative techno-economic analysis of hybrid micro-grid systems utilizing different battery types. *Energy Conversion and Management*, 2016, 112, 435–44. doi:10.1016/j.enconman.2016.01.014.

[18] International Electrotechnical Commission (IEC). *Electrical Energy Storage White Paper.*, vol. IEC WP EES. 2011.

[19] Lawder, MT; Suthar, B; Northrop, PWC; De, S; Hoff, CM; Leitermann, O; et al. Battery energy storage system (BESS) and battery management system (BMS) for grid-scale applications. *Proceedings of the IEEE*, 2014, 102, 1014–30. doi:10.1109/JPROC.2014.2317451.

[20] Divya, KC; Østergaard, J. Battery energy storage technology for power systems-An overview. *Electric Power Systems Research*, 2009, 79, 511–20. doi:10.1016/j.epsr.2008.09.017.

[21] Castillo, A; Gayme, DF. Grid-scale energy storage applications in renewable energy integration: A survey. *Energy Conversion and Management*, 2014, 87, 885–94. doi:10.1016/j.enconman.2014.07.063.

[22] Wade, NS; Taylor, PC; Lang, PD; Jones, PR. Evaluating the benefits of an electrical energy storage system in a future smart grid. *Energy Policy*, 2010, 38, 7180–8. doi:10.1016/j.enpol.2010.07.045.

[23] Cho, J; Kleit, AN. Energy storage systems in energy and ancillary markets: A backwards induction approach. *Applied Energy*, 2015, 147, 176–83. doi:10.1016/j.apenergy.2015.01.114.

[24] Dufo-López, R; Bernal-Agustín, JL. Techno-economic analysis of grid-connected battery storage. *Energy Conversion and Management*, 2015, 91, 394–404. doi:10.1016/j.enconman.2014.12.038.

[25] Consonni, S; Giugliano, M; Grosso, M. Alternative strategies for energy recovery from municipal solid waste: Part B: Emission and cost estimates. *Waste Management*, 2005, 25, 137–48. doi:10.1016/j.wasman.2004.09.006.

[26] Chua, KH; Lim, YS; Morris, S. Cost-benefit assessment of energy storage for utility and customers: A case study in Malaysia. *Energy Conversion and Management*, 2015, 106, 1071–81. doi:10.1016/j.enconman.2015.10.041.

[27] Schrøder Pedersen A. European Energy Storage Technology Development Roadmap towards 2030. *International Energy Storage Policy and Regulation Workshop*, 2014.

[28] World Energy Council. E-storage: Shifting from cost to value. Wind and solar applications. *World Future Energy Summit*, 2016, p. 1–14.

[29] International Renewable Energy Agency (IRENA). *Renewables and Electricity Storage - A technology roadmap for REmap 2030.*, 2015. doi:10.1007/978-88-470-1998-0_8.

[30] Zucker, A; Hinchliffe, T; Spisto, A. Assessing storage value in electricity markets a literature review. *JRC Scientific and Policy Reports European Union*, 2013, 74. doi:10.2790/89242.

[31] Khajesalehi, J; Hamzeh, M; Sheshyekani, K; Afjei, E. Modeling and control of quasi Z-source inverters for parallel operation of battery energy storage systems: Application to microgrids. *Electric Power Systems Research*, 2015, 125, 164–73.

[32] Fernão Pires, V; Romero-Cadaval, E; Vinnikov, D; Roasto, I; Martins, JF. Power converter interfaces for electrochemical energy storage systems - A review. *Energy Conversion and Management*, 2014, 86, 453–75. doi:10.1016/j.enconman.2014.05.003.

[33] Harkare, C; Kapoor, P. Study and Simulation of Current Controlled PWM Inverters and their applications. *International Journal on Recent and Innovation Trends in Computing and Communication*, 2015, 3, 138–42.

[34] Mesa, L; Jimena, D; Muñoz, C; Alberto, G; Chávez, D; Oscar, J; et al. *Modulación PWM aplicada a inversores trifásicos dentro del esquema de accionamientos eléctricos AC* [*PWM modulation applied to three-phase inverters within the scheme of AC electric drives*]. Primer ISA Show Andino 2007 Ponencia, 2007, p. 1–24.

[35] Kazmierkowski, MP; Malesani, L. Current control techniques for three-phase voltage-source PWM\nconverters: a survey. *IEEE*

Transactions on Industrial Electronics, 1998, 45, 691–703. doi:10.1109/41.720325.

[36] Kim, HS; Ryu, MH; Baek, JW; Jung, JH. High-efficiency isolated bidirectional AC-DC converter for a DC distribution system. *IEEE Transactions on Power Electronics*, 2013, 28, 1642–54. doi:10.1109/TPEL.2012.2213347.

[37] Wu, TF; Chang, CH; Lin, LC; Yu, GR; Chang, YR. DC-bus voltage control with a three-phase bidirectional inverter for DC distribution systems. *IEEE Transactions on Power Electronics*, 2013, 28, 1890–9. doi:10.1109/TPEL.2012.2206057.

[38] Wu, TF; Kuo, CL; Lin, LC; Chen, YK. DC-Bus Voltage Regulation for a DC Distribution System With a Single-Phase Bidirectional Inverter. *IEEE Journal of Emerging and Selected Topics in Power Electronics*, 2016, 4, 210–20. doi:10.1109/JESTPE.2015.2485300.

[39] Ganesan, SI; Pattabiraman, D; Govindarajan, RK; Rajan, M; Nagamani, C. Control Scheme for a Bidirectional Converter in a Self-Sustaining Low-Voltage DC Nanogrid. *IEEE Transactions on Industrial Electronics*, 2015, 62, 6317–26. doi:10.1109/TIE.2015.2424192.

[40] Caruana, C; Sattar, A; Al-Durra, A; Muyeen, SM. Real-time testing of energy storage systems in renewable energy applications. *Sustainable Energy Technologies and Assessments*, 2015, 12, 1–9.

[41] Urtasun, A; Sanchis, P; Marroyo, L. State-of-charge-based droop control for stand-alone AC supply systems with distributed energy storage. *Energy Conversion and Management*, 2015, 106, 709–20. doi:10.1016/j.enconman.2015.10.010.

[42] Akhil, AA; Huff, G; Currier, AB; Kaun, BC; Rastler, DM; Chen, SB; et al. *DOE/EPRI Electricity Storage Handbook in Collaboration with NRECA.*, 2015.

[43] Reihani, E; Sepasi, S; Roose, LR; Matsuura, M. Energy management at the distribution grid using a Battery Energy Storage System (BESS). *International Journal of Electrical Power & Energy Systems*, 2016, 77, 337–44. doi:10.1016/j.ijepes.2015.11.035.

[44] Ginart, A; Salazar, A; Liou, R. Transformerless Bidirectional Inverter for Residential Battery Storage Systems. *IEEE Green Technologies Conference (GreenTech)*, Kansas City, MO, 2016, p. 18–23. doi:10.1109/GreenTech.2016.11.

[45] Teymour, HR; Sutanto, D; Muttaqi, KM; Ciufo, P. Solar PV and battery storage integration using a new configuration of a three-level NPC inverter with advanced control strategy. *IEEE Transactions on Energy Conversion*, 2014, 29, 354–65. doi:10.1109/TEC.2014.2309698.

[46] Jou, HL; Chang, YH, Wu J-C, Wu K-D. Operation strategy for a lab-scale grid-connected photovoltaic generation system integrated with battery energy storage. *Energy Conversion and Management*, 2015, 89, 197–204.

[47] Waltrich, G; Duarte, JL; Hendrix, MAM. Three-phase bidirectional dc/ac converter using a six-leg inverter connected to a direct ac/ac converter. *IET Power Electronics*, 2015, 8, 2214–22. doi:10.1049/iet-pel.2015.0059.

[48] Kumar, A; Neogi, N. Bidirectional Converter and Energy Storage System. *International Journal of Enhanced Research in Science Technology & Engineering*, 2015, 4, 15–23.

[49] Everts, J; Krismer, F; Van Den Keybus, J; Driesen, J; Kolar, JW. Optimal ZVS Modulation of Single-Phase Single-Stage Bidirectional DAB AC-DC converters. *IEEE Transactions on Power Electronics*, 2014, 29, 3954–70. doi:10.1109/TPEL.2013.2292026.

[50] Danyali, S; Mozaffari Niapour, SAK; Hosseini, SH; Gharehpetian, GB; Sabahi, M. New Single-Stage Single-Phase Three-Input DC-AC Boost Converter for Stand-Alone Hybrid PV/FC/UC Systems. *Electric Power Systems Research*, 2015, 127, 1–12. doi:10.1016/j.epsr.2015.05.008.

[51] Bhattacharya, S; Deb, P; Biswas, SK; KarChowdhury, S. Performance and design of an open-delta connected grid tied bidirectional PWM converter. *International Journal of Electrical Power and Energy Systems*, 2016, 78, 183–93. doi:10.1016/j.ijepes.2015.11.079.

[52] Singh, B; Singh, BN; Chandra, A; Al-Haddad, K; Pandey, A; Kothari, DP. A review of three-phase improved power quality AC-DC converters. *IEEE Transactions on Industrial Electronics*, 2004, 51, 641–60. doi:10.1109/TIE.2004.825341.

[53] MathWorks®. *MATLAB® Simulink® 2016*. http://www.mathworks.com/(accessed 18 January 2016).

[54] Zeng, Z; Yang, H; Zhao, R; Cheng, C. Topologies and control strategies of multi-functional grid-connected inverters for power quality enhancement: A comprehensive review. *Renewable and Sustainable Energy Reviews*, 2013, 24, 223–70. doi:10.1016/j.rser.2013.03.033.

[55] Reddy, BV; Babu, BC. Hysteresis controller and delta modulator- A two viable scheme for current controlled voltage Source Inverter. *International Conference for Technical Postgraduates*, 2009, TECHPOS 2009 2009. doi:10.1109/TECHPOS.2009.5412058.

[56] Malesani, L; Tomasin, P. PWM current control techniques of voltage source converters-a survey. *Proceedings of IECON '93 - 19th Annual Conference of IEEE Industrial Electronics*, 1993, p. 670–5. doi:10.1109/IECON.1993.339000.

[57] Mohapatra, M; Babu, BC. Fixed and sinusoidal-band hysteresis current controller for PWM voltage source inverter with LC filter. *TechSym 2010 - Proceedings of the 2010 IEEE Students' Technology Symposium*, 2010, 88–93. doi:10.1109/TECHSYM.2010.5469205.

[58] Kalyanraj, D; Lenin Prakash, S. Design and Digital Implementation of Constant Frequency Hysteresis Current Controller for Three-Phase Voltage Source Inverter Using TMS320F2812. *International Journal of Emerging Electric Power Systems*, 2014, 15, 13–23. doi:10.1515/ijeeps-2013-0141.

[59] Brod, DM; Novotny, DW. Current Control of VSI-PWM Inverters. *IEEE Transactions on Industry Applications*, 1985, IA-21:562–70. doi:10.1109/TIA.1985.349711.

[60] Karris, S. Introduction to Simulink® with engineering applications. *Mathematics*, 2006, p. 572.

[61] *European Commission. Impact Assessment Guidelines*. vol. 92, SEC., 2009.

[62] European Commission. Joint Research Centre (JRC). *Sensitivity Analysis*, 2016. http://sensitivity-analysis.jrc.ec.europa.eu (accessed 10 April 2016).

[63] Saltelli, A; Ratto, M; Andres, T; Campolongo, F; Cariboni, J; Gatelli, D; et al. *Global sensitivity analysis: the primer*, 2008. doi:10.1002/9780470725184.

[64] Rampinelli, GA; Gasparin, FP; Bühler, AJ; Krenzinger, A; Chenlo Romero, F. Assessment and mathematical modeling of energy quality parameters of grid connected photovoltaic inverters. *Renewable and Sustainable Energy Reviews*, 2015, 52, 133–41. doi:10.1016/j.rser. 2015.07.087.

[65] Rastler, DM. Electricity Energy Storage Technology Options. A White Paper Primer on Applications, Costs, and Benefits. *Electric Power Research Institute (EPRI)*, 2010, 1020676, 1–170.

About the Editors

Antonio Colmenar Santos, PhD

Full Professor
Universidad Nacional de Educación A Distancia, Spain
Email: acolmenar@ieec.uned.es

Dr. Antonio Colmenar-Santos has been a senior lecturer in the field of Electrical Engineering at the Department of Electrical, Electronic and Control Engineering at the National Distance Education University (UNED) since June 2014. Dr. Colmenar-Santos was an adjunct lecturer at both the Department of Electronic Technology at the University of Alcalá and at the Department of Electric, Electronic and Control Engineering at UNED. He has also worked as a consultant for the INTECNA project (Nicaragua). He has been part of the Spanish section of the International Solar Energy Society (ISES) and of the Association for the Advancement of Computing in Education (AACE), working in a number of projects related to renewable energies and multimedia systems applied to teaching. He was the coordinator of both the virtualisation and telematic Services at ETSII-UNED, and deputy head teacher and the head of the Department of Electrical, Electronics and Control Engineering at UNED. He is the author of more than 50 papers published in respected journals (http://goo.gl/YqvYLk) and has participated in more than 100 national and international conferences.

Enrique Rosales Asensio, PhD

Researcher
Universidad de La Laguna, Spain
Email: erosalea@ull.edu.es

Dr. Enrique Rosales-Asensio is an industrial engineer with postgraduate degrees in electrical engineering, business administration, and quality, health, safety and environment management systems. Currently, he is a senior researcher at the University of La Laguna, where he is involved in water desalination project in which the resulting surplus electricity and water would be sold. He has also worked as a plant engineer for a company that focuses on the design, development and manufacture of waste-heat-recovery technology for large reciprocating engines; and as a project manager in a world-leading research centre.

David Borge Diez, PhD

Lecturer
Universidad de León, Spain
Email: dbord@unileon.es

Dr. David Borge-Diez has a PhD in Industrial Engineering and an MSc in Industrial Engineering, both from the School of Industrial Engineering at the National Distance Education University (UNED). He is currently a lecturer and researcher at the Department of Electrical, Systems and Control Engineering at the University of León, Spain. He has been involved in many national and international research projects investigating energy efficiency and renewable energies. He has also worked in Spanish and international engineering companies in the field of energy efficiency and renewable energy for over eight years. He has authored more than 25 publications in international peer-reviewed research journals and participated in numerous international conferences.

INDEX

A

B

C

D

E

F

G

H

I

J

L

M

N

O

P

Q

R

S

T

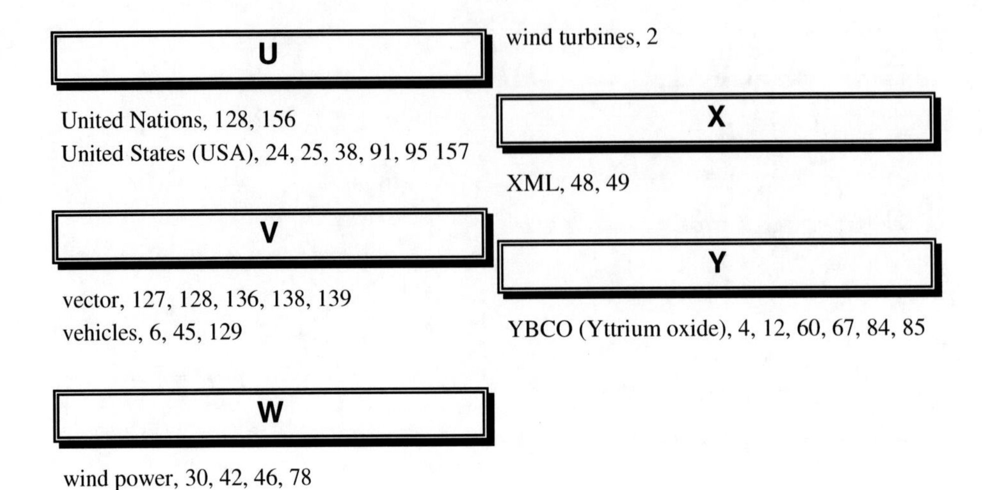

U

V

W

X

Y

Focus on Renewable Energy Sources

Editors: Giuseppe Scandurra (University of Naples "Parthenope" Naples, Italy)

Series: Renewable Energy: Research, Development and Policies

Book Description: With this volume, the authors want to explore and analyze the causes and consequences of fragmentation and discussing policy responses on promoting renewable energy generation by shedding light on the policies proposed to promote the renewable generation and enhance energy efficiency.

Hardcover ISBN: 978-1-53613-802-3
Retail Price: $160

Bioethanol and Beyond: Advances in Production Process and Future Directions

Editor: Michel Brienzo (Laboratory of Macromolecules, Division of Metrology Applied to life Science – National Institute of Metrology, Quality and Technology – INMETRO, Duque de Caxias – RJ, Brazil.)

Series: Renewable Energy: Research, Development and Policies

Book Description: *Bioethanol and Beyond: Advances in Production Process and Future Directions* discusses the advances in the production process of bioethanol, ranging from first to fourth generation bioethanol.

Hardcover ISBN: 978-1-53613-478-0
Retail Price: $270